Ben Stacy Jerrik (Ed.)

Denali Highway

AF307146

Ben Stacy Jerrik (Ed.)

Denali Highway

Richardson Highway, Denali National Park and Preserve, George Parks Highway, Tangle Lakes

Part Press

Imprint

Permission is granted to copy, distribute and/or modify this document under the terms of the GNU Free Documentation License, Version 1.2 or any later version published by the Free Software Foundation; with no Invariant Sections, with the Front-Cover Texts, and with the Back- Cover Texts. A copy of the license is included in the section entitled "GNU Free Documentation License".

All parts of this book are extracted from Wikipedia, the free encyclopedia (www.wikipedia.org).

You can get detailed informations about the authors of this collection of articles at the end of this book. The editors (Ed.) of this book are no authors. They have not modified or extended the original texts.

Pictures published in this book can be under different licences than the GNU Free Documentation License. You can get detailed informations about the authors and licences of pictures at the end of this book.

The content of this book was generated collaboratively by volunteers. Please be advised that nothing found here has necessarily been reviewed by people with the expertise required to provide you with complete, accurate or reliable information. Some information in this book maybe misleading or wrong. The Publisher does not guarantee the validity of the information found here. If you need specific advice (f.e. in fields of medical, legal, financial, or risk management questions) please contact a professional who is licensed or knowledgeable in that area.

Any brand names and product names mentioned in this book are subject to trademark, brand or patent protection and are trademarks or registered trademarks of their respective holders. The use of brand names, product names, common names, trade names, product descriptions etc. even without a particular marking in this works is in no way to be construed to mean that such names may be regarded as unrestricted in respect of trademark and brand protection legislation and could thus be used by anyone.

Cover image: www.ingimage.com
Concerning the licence of the cover image please contact ingimage.

Publisher:
Part Press is a trademark of
International Book Market Service Ltd., 17 Rue Meldrum, Beau Bassin, 1713-01 Mauritius
Email: info@bookmarketservice.com
Website: www.bookmarketservice.com

Published in 2011

Printed in: U.S.A., U.K., Germany. This book was not produced in Mauritius.

ISBN: 978-613-8-66159-7

Contents

Denali Highway

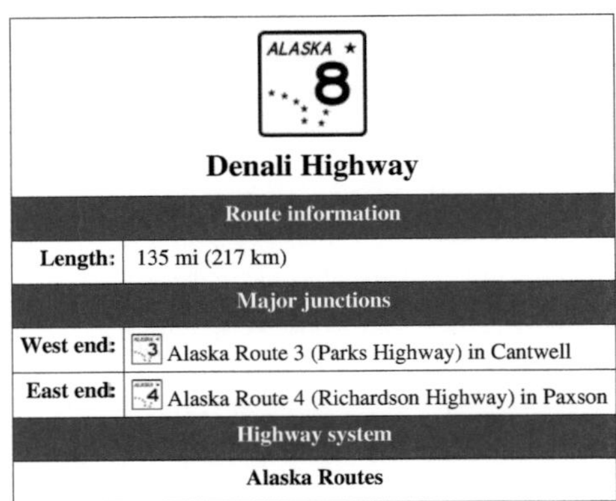

Denali Highway

Route information	
Length:	135 mi (217 km)
Major junctions	
West end:	Alaska Route 3 (Parks Highway) in Cantwell
East end:	Alaska Route 4 (Richardson Highway) in Paxson
Highway system	
Alaska Routes	

Denali Highway (Alaska Route 8) is a lightly traveled, mostly gravel highway in the U.S. state of Alaska. It leads from Paxson on the Richardson Highway to Cantwell on the Parks Highway. Opened in 1957, it was the first road access to Denali National Park (then known as Mount McKinley National Park). Since 1971, primary park access has been via the Parks Highway, which incorporated a section of the Denali Highway from Cantwell to the present-day park entrance. The Denali Highway is 135 miles (217 km) in length.

Conditions

The highway is now little used and poorly maintained, and closed to all traffic from October to mid-May each year. Only the easternmost 21.3 miles (34.3 km) and westernmost 2.6 miles (4.2 km) are paved; whether the remainder should be paved as well is a continual source of debate. Washboarding and extreme dust are common, the recommended speed limit is 30 mph (48 km/h).

Route description

The Denali Highway is seen in summer.

Traveling west, the Denali Highway leaves the Richardson Highway (Alaska Route 4) at Paxson, and climbs steeply up into the foothills of the central Alaska Range. The first 21 miles (34 km), to Tangle Lakes, are paved. Along its length, the highway passes through three of the principal river drainages in Interior Alaska: the Copper River drainage, the Tanana/Yukon drainage and the Susitna drainage. Along the way, in good weather, there are stunning views of the peaks and glaciers of the central Alaska Range, including Mount Hayes (13,700 ft), Mount Hess (11,940 ft) and Mount Deborah (12,688 ft). At MP 15, from the pullout on the south side of the road, in clear weather you can see the Wrangell Mountains, the Chugach Mountains and the Alaska Range.

The first 45 miles (72 km) winds through the Amphitheater Mountains, cresting at Maclaren Summit, at 4086 feet (1245 m) the second highest road in Alaska. The road then drops down to the Maclaren River Valley with fine views north to Maclaren Glacier. After crossing the Maclaren River, the road winds through the geologically mysterious Crazy Notch and then along the toe of the Denali Clearwater Mountains to the Susitna River. After crossing the Susitna River the road extends across the glaciers outwash plains to the Nenana River, and then down the Nenana River to Cantwell on the George Parks Highway (Alaska #3).

Services

There are developed campgrounds at Tangle Lakes (MP 22) and Brushkana Creek (MP 104), but there are dozens of pullouts where you can camp on public lands.

Services are scant along this road. Year-round operations include Denali Highway Cabins & Tours (MP 0.2), Maclaren River Lodge (MP 42), Alpine Creek Lodge (MP68) and Backwoods Lodge (MP134); summer-only operations include Tangle River Inn (MP 20) and Gracious House (MP 82) - limited service there summer 2010. Winter travel on the Denali Highway is exclusively by snowmobile and dogsled. Automobile travelers are severely discouraged from attempting to traverse the road in winter; as recently as 1996 three persons died from exposure when snows blocked their progress. The road is cleared by DOT late in April and generally is passable by non-4WD from then until the first snows close it, usually late September on the eastern, tundra end and late October-early November on the lower, boreal forest western end.

Recreation

The Tangle Lakes constitute the headwaters of the Delta River, a popular destination for canoeists as it is the launch point of the Delta River Canoe Trail.[1]

The Denali Highway is an important birding destination. It offers road access to alpine terrain — not that common in Alaska — and, in the brief birding season there, good viewing of a number of alpine breeders, including Arctic Warbler, Smith's Longspur, Long-tailed Jaeger, Whimbrel, Surfbird, Lapland Longspur, Horned Lark, Short-eared Owl, Wandering Tattler, Gyrfalcon and much more. A walk north along BLM's Maclaren Summit Trail (MP 39) can be very productive. There are also trumpeter swans and various other waterfowl in the lakes and ponds along the route.

White Spruce taiga along the Denali Highway, with the Alaska Mountain Range in the Background

Fishing for grayling and lake trout is decent, if not spectacular, in any of the clear water (i.e., unglaciated) streams.

Because the area is hunted heavily, larger mammals are much less common than in Denali National Park, but moose, grizzly bear, and caribou are fairly common. The Nelchina caribou herd, approximately 36,000 animals as of winter 2009–2010, normally passes through this area after calving season ends, and some autumns and winters as many as 16,000 animals can be seen at once. The herd forms an important foodsource for many residents of southcentral Alaska, and visitors eager to view the animals may be competing with hunters. The many lakes along the road are also a destination for duck hunting in the fall.

Most of the land along the highway is publicly owned. There are several BLM-maintained trails, and dozens of informal trails. This is a stretch of wild Alaska that is pretty much unspoiled, relatively accessible and beautiful.

Gallery

At 4086 feet (1245 m), MacClaren summit is the highest point on the highway.

Portions of the road are built directly on top of glacial eskers.

On a clear day, westbound travelers can see Mt. McKinley

Numerous kettle lakes line the eastern portion of the highway.

The Susitna River crossing is the only large bridge on the highway.

See also

- List of Alaska Routes

Further reading

- United States. Bureau of Land Management. Glennallen Field Office. (2007). *Denali Highway: points of interest* [BLM Recreation Guide BLM/AK/GI-88/023+8351+050, Rev. 07]. Glennallen, AK: author.

External links

- Cycling the Denali Highway, including altitude profiles, on WorldOnaBike.com [2]
- The Denali Highway spring travelogue [3]
- The Denali Highway autumn travelogue [4]
- Driving the Denali Highway [5]

References

[1] The Milepost 59th edition (2007) ISBN 1-892154-21-8 page 497

[2] http://worldonabike.com/trip-reports/north-america/alaska/
     day-27-30-7-10-aug-2008-the-denali-highway-about-rough-roads-cold-and-more/

[3] http://members.virtualtourist.com/m/tt/8fac3/#TL

[4] http://members.virtualtourist.com/m/tt/95fd3/#TL

[5] http://www.icdc.com/~neubauer/denali.htm

Richardson Highway

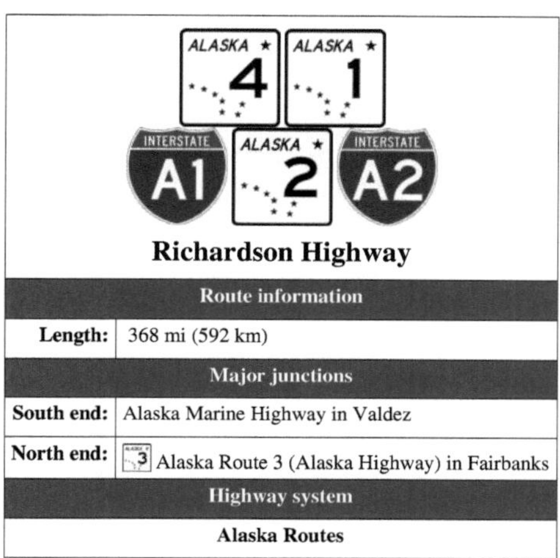

Richardson Highway

Route information	
Length:	368 mi (592 km)
Major junctions	
South end:	Alaska Marine Highway in Valdez
North end:	Alaska Route 3 (Alaska Highway) in Fairbanks
Highway system	
Alaska Routes	

The **Richardson Highway** is a highway in the U.S. state of Alaska, running 368 miles (562 km) from Valdez to Fairbanks. It is marked as **Alaska Route 4** from Valdez to Delta Junction and as Alaska Route 2 from there to Fairbanks. It is also connects segments of Alaska Route 1 between the Glenn Highway and the Tok Cut-Off. It was the first major road built in Alaska.

History

A pack trail from the port at Valdez to Eagle, a distance of about 409 miles (660 km), was built in 1898 by the U.S. Army to provide an "all-American" route to the Klondike gold fields. After the rush ended, the Army kept the trail open in order to connect its posts at Fort Liscum, in Valdez, and Fort Egbert, in Eagle.

"first car" to travel from Valdez to Eagle, 1913

The Fairbanks gold rush in 1902, and the construction of a WAMCATS telegraph line along the trail in 1903, made the Valdez-to-Eagle trail one of the most important access routes to the Alaska Interior, so in 1910, the Alaska Road Commission upgraded it to a wagon road. The head of the project was U.S. Army General Wilds P. Richardson, for whom the highway was later named. During the construction, the government hired failed gold prospectors as well as regular construction workers. The income from this work allowed many of the prospectors to leave Alaska. Several roadhouses now on the National Register of Historic Places were constructed along the route at this time.

Richardson Highway from the air, cutting behind a meadow lake.

The rise of motorized travel led the road to be upgraded to automobile standards in the 1920s. To finance continued maintenance and road construction, the Alaska Road Commission instituted tolls for commercial vehicles in 1933 of up to $175 per trip, which were collected at the Tanana River ferry crossing at Big Delta. When the tolls were further increased in 1941 to boost business for the Alaska Railroad, disgruntled truckers nicknamed "gypsies" started a rogue ferry service in order to evade the toll.

The Alaska and Glenn highways, built during World War II, connected the rest of the continent and Anchorage to the Richardson Highway at Delta Junction and Glennallen respectively, allowing motor access to the new military bases built in the Territory just prior to the war. The bridge at Big Delta, the last remaining gap, was built as part of the Alaska Highway project.

The southern end was only open during summers until 1950, when a freight company foreman who lived near the treacherous Thompson Pass plowed the snow himself for an entire season to prove the route could be used year-round. The highway was paved in 1957.

The Trans-Alaska Pipeline System, built in 1973-1977, mostly parallels the highway from Fairbanks to Valdez.

Recent and future improvements

- During the 1990s, the highway was upgraded from Fairbanks to the main gate at Eielson AFB, making this stretch a 4-lane divided road. Intersections with other roads, however, are still almost entirely at-grade.

- Under SAFETEA-LU, Alaska Route 2 from the Canadian border to Fairbanks, comprising parts of the Richardson and Alaska Highways, has been declared a High Priority Corridor (Corridor 67). What this means for the distant future is not yet certain; although SAFETEA-LU does explicitly provide federal funds for upgrading the road to 4 lanes and divided, from Salcha to Delta Junction.

Interchange with the Fairbanks end of Badger Road, a loop road serving suburban parts of the North Pole area. The interchange was constructed in the early 2000s, after an approximately three-decade history of serious accidents at the previous at-grade intersection.

Towns and places along the Richardson Highway

- Valdez, no milepost (*The Milepost* lists it as 4 miles (6.4 km) from Mile 0)
- Old Valdez (destroyed in Good Friday Earthquake), mile 0 (km 0)
- Copper Center, mile 100 (km 162)
- Glennallen (Glenn Highway), mile 115 (km 185)
- Gulkana, mile 127 (km 204)
- Gakona Junction (Tok Cut-Off), mile 129 (km 207)
- Paxson (Denali Highway), mile 186 (km 299)
- Isabel Pass, where the highway crosses the Alaska Range
- Black Rapids Roadhouse
- Fort Greely, mile 261 (km 420)

A buried crossing of the highway by the oil pipeline.

- Delta Junction (Alaska Highway), mile 266 (km 428)
- Big Delta, Alaska and Rika's Landing Roadhouse, mile 275
- Salcha, mile 325 (km 524)
- Eielson Air Force Base, mile 341 (km 549)

- North Pole, mile 349 (km 562)
- Fairbanks, mile 364 (km 586)

External links

- Evolution of the Richardson Highway - ExploreNorth [1]
- A journey down the Richardson Highway [2]

References

[1] http://www.explorenorth.com/library/yafeatures/bl-richardson.htm

[2] http://members.virtualtourist.com/m/tt/8fe31/#TL

Denali National Park and Preserve

<table>
<tr><td colspan="2" align="center">Denali National Park and Preserve</td></tr>
<tr><td colspan="2" align="center">IUCN Category II (National Park)</td></tr>
<tr><td>Location</td><td>Denali Borough and Matanuska-Susitna Borough, Alaska, USA</td></tr>
<tr><td>Nearest city</td><td>Healy</td></tr>
<tr><td>Coordinates</td><td>63°20′0″N 150°30′0″W</td></tr>
<tr><td>Area</td><td>6075107 acres (24585.09 km^2)</td></tr>
<tr><td>Visitors</td><td>1,178,745 (in 2005)</td></tr>
</table>

Denali National Park and Preserve is located in Interior Alaska and contains Denali (Mount McKinley), the highest mountain in North America. The park and preserve together cover 9,492 mi² (24,585 km²).The longest glacier is the Kalhiltna glacier.

Overview

The word "Denali" means "the high one" in the native Athabaskan language and refers to the mountain itself. The mountain was named after president William McKinley of Ohio in 1897 by local prospector William A. Dickey, although McKinley had no connection with the region. The name is only used by those outside of Alaska. Charles Alexander Sheldon took an interest in the Dall sheep native to the region, and became concerned that human encroachment might threaten the species. After his 1907-1908 visit, he petitioned the people of Alaska and Congress to create a preserve for the sheep. (His account of the visit was published posthumously as *The Wilderness of Denali*, ISBN 1-56833-152-5). The park was established as **Mount McKinley National Park** on February 26, 1917. However, only a portion of Mount McKinley (not even including the summit) was within the original park boundary. The park was designated an international biosphere reserve in 1976. A separate **Denali National Monument** was proclaimed by Jimmy Carter on December 1, 1978.

Mount McKinley National Park, whose name had been subject to local criticism from the onset, and Denali National Monument were incorporated and established into **Denali National Park and Preserve** by the Alaska National Interest Lands Conservation Act, December 2, 1980. At this time the Alaska Board of Geographic Names changed the name of the mountain back to "Denali," even though the U.S. Board of Geographic Names maintains "McKinley". Alaskans tend to use "Denali" and rely on context to distinguish between the park and the mountain. The size of the national park is over 6 million acres (24,500 km²), of which 4,724,735.16 acres (19,120 km²) are federally owned. The national preserve is 1,334,200 acres (543 km²), of which 1,304,132 acres (5,278 km²) are federally owned. On December 2, 1980, a 2,146,580 acre (8,687 km²) Denali Wilderness was established within the park.

Denali as seen from the park road

Denali habitat is a mix of forest at the lowest elevations, including deciduous taiga. The preserve is also home to tundra at middle elevations, and glaciers, rock, and snow at the highest elevations. Today, the park hosts more than 400,000 visitors who enjoy wildlife viewing, mountaineering, and backpacking. Wintertime recreation includes dog-sledding, cross-country skiing, and snowmobiling where allowed.

Denali National Park

Wildlife

Denali is home to a variety of Alaskan birds and mammals, including a healthy population of grizzly bears and black bears. Herds of caribou roam throughout the park. Dall sheep are often seen on mountainsides, and moose feed on the aquatic plants of the small lakes and swamps. Despite human impact on the area, Denali accommodates gray wolf dens, both historic and active. Smaller animals, such as hoary marmots, arctic ground squirrels, beavers, pikas, and snowshoe hares are seen in abundance. Foxes, martens, lynx, wolverines also inhabit the park, but are more rarely seen due to their elusive natures.

Grizzly bear in Denali National Park

The park is also well known for its bird population. Many migratory species reside in the park during late spring and summer. Birdwatchers may find waxwings, Arctic Warblers, pine grosbeaks, and wheatears, as well as Ptarmigan and the majestic tundra swan. Predatory birds include a variety of hawks, owls, and the gyrfalcon, as well as the abundant but striking golden eagle.

Ten species of fish, including trout, salmon, and arctic grayling share the waters of the park. Because many of the rivers and lakes of Denali are fed by glaciers, glacial silt and cold temperatures slow the metabolism of the fish, preventing them from reaching normal sizes. A single amphibious species, the wood frog, also lives among the lakes of the park.

Denali park rangers maintain a constant effort to keep the wildlife wild by limiting the interaction between humans and park animals. However, the number of wild bears necessitates their wearing collars to track movements. Feeding animals is strictly forbidden, as it may cause adverse affects on the feeding habits of the creature. Visitors are encouraged to view animals from safe distances. Despite the large concentration of bears in the park, efforts by rangers to educate backpackers and visitors about preventative measures and BRFCs have greatly reduced the number of dangerous encounters. Certain areas of the park are often closed due to uncommon wildlife activity, such as denning areas of wolves and bears or recent kill sites. These restricted areas may change throughout the year. Through the collective care of park staff and visitors, Denali has become a premier destination for wildlife viewing.

Geography

The Alaska Range, a mountainous expanse running through the entire park, provides interesting ecosystems in Denali. Because the fall line lies as low as 2500 feet (760 m), wooded areas are rare inside the park, except in the flatter western sections surrounding Wonder Lake, most of the park is vast expanses of tundra. and lowlands of the park where flowing waters melt the frozen ground. Spruces and willows make up the majority of these treed areas. Because of mineral content, ground temperature, and a general lack of soil, areas surrounding the bases of mountains are not suitable for sufficient tree growth, and most trees and shrubs do not reach full size.

Painting of the heavily glaciated southern part of Denali, by Heinrich Berann

Having a range of elevations, there is a variety of vegetation zones. From lowest to highest, there is low brush bog, bottomland spruce-poplar forest, upland spruce-hardwood forest, moist tundra, and finally the highest of elevations, alpine tundra.

Throughout Denali's history, there has been a patchwork pattern of different plants relying on fire. Because of this, the fire history is too complicated to explain. North of the Alaskan Range, fires are common, occurring when old forests need replacement.

Tundra is the predominate ground cover of the park. Layers of topsoil collect on rotten, fragmented rock moved by thousands of years of glacial activity. Mosses, ferns, grasses, and fungi quickly fill the topsoil, and in areas of muskeg "wet tundra," tussocks form and may

Alpine forest and lakes in Denali by Carol M. Highsmith

collect algae. The term muskeg includes very spongy waterlogged tussocks as well as deep pools of water covered by solid looking moss. Wild blueberries and soap berries thrive in this landscape, and provide the bears of Denali with the main part of their diets.

Over 450 species of flowering plants fill the park, and can be viewed in bloom throughout summer. Images of goldenrod, fireweed, lupine, bluebell. and gentian filling the valleys of Denali are often used on postcards and in artwork.

Climate

Long winters are followed by short growing seasons. Eighty percent of the bird population returns after cold months, raising their young. Most mammals and other wildlife in the park spend the brief summer months preparing for winter and raising their young.

Summers are usually cool and damp, but temperatures in the 70s are not rare. The weather is so unpredictable that there have even been instances of snow in August.

The north and south side of the Alaskan Range have a completely different climate. The Gulf of Alaska carries moisture to the south side, but the mountains block water to the north side. This brings a drier climate and huge temperature fluctuations to the north. The south receives transitional maritime continental climates, with moister, cooler summers and warmer winters.

Vehicle access

The park is serviced by a 91-mile (146 km) road from the George Parks Highway to the mining camp of Kantishna. It runs east to west, north of and roughly parallel to the imposing Alaska Range. Only a small fraction of the road is paved because permafrost and the freeze-thaw cycle create an enormous cost for maintaining the road. Only the first 15 miles (24 km) of the road are available to private vehicles, and beyond this point visitors must access the interior of the park through concessionary buses. Wonder Lake can be reached by a six-hour bus ride from the Wilderness Access Center. Eielson Visitor Center is located four hours into the park on the road.

The single road within Denali National Park

Several fully narrated tours of the park are available, the most popular of which is the Tundra Wilderness Tour. The tours travel from the initial boreal forests through tundra to the Toklat River or Kantishna. A clear view of the mountain is only possible about 20% of the time during the summer, although it is visible more often during the winter. Several portions of the road run alongside sheer cliffs that drop hundreds of feet at the edges, and there are no guardrails. As a result of the danger involved, and because most of the gravel road is only one lane wide, drivers are trained extensively in procedures for navigating the sharp mountain curves, and yielding the right-of-way to opposing buses and park vehicles.

While the main park road goes straight through the middle of the Denali National Park Wilderness, the national preserve and portions of the park not designated wilderness are even more inaccessible. There are no roads extending out to the preserve areas, which are on the far west end of the park. The far north of the park, characterized by hills and rivers, is accessed by the Stampede Trail, a dirt road which stops at the park boundary. The very rugged south portion of the park, characterized by enormous glacier-filled canyons, is accessed by Petersville Road, a dirt road that stops about 5 miles (8.0 km) outside of the park. The mountains can be accessed most easily by air taxis that land on the glaciers.

Wilderness

The Denali Wilderness is a wilderness area in the Denali National Park and Preserve. It encompasses the high heart of the Alaska Range, including Denali, the centerpiece of the wilderness, which comprises about one-third of the national park.

Denali Wilderness covers the area formerly known as Mount McKinley National Park from 1917 until the park was expanded and renamed in 1980. It is 2,146,580 acres (8,687 km²) in area; the entire park is larger than the state of Massachusetts.[1]

Prehistory and protohistory

An immense collection of cultural sites gives more and more clues as to what and who used to live there. Thousand of years ago, grassland was abundant, and mammoths utilized the flat *Mammoth Steppe* to move and graze. Around 11,000 to 13,500 years ago, these grasslands shrunk and woody shrubs began to appear. Back then, the North-Alaskan Range was predominantly ice free.

187 cultural sites tell about Denali's past, eighty-four have prehistoric items. Native Americans have lived in this environment for 11,000 years, using every resource the wild provided. The Koyukon, Dena'ina, Athna, Kolchan, Tanana, and Athabaskans are particularly known.

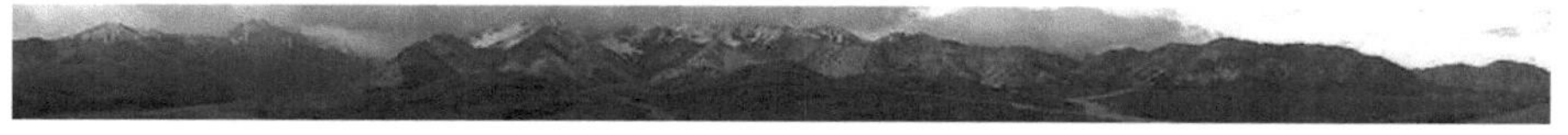

Panoramic view of the Polychrome Mountains

Fossils

Denali is emerging as a site of interesting fossils, including footprints (ichnites) that were credited with being the first evidence of prehistoric wading birds, probing in mudflats for food during the Late Cretaceous Period, when they were first publicly reported in May 2006. A footprint of a carnivorous theropod had previously been reported in the park.

References

[1] Wilderness.net – Denali Wilderness (http://www.wilderness.net/index.cfm?fuse=NWPS&sec=wildView&wname=Denali)

External links

- National Park Service: Denali National Park & Preserve (http://www.nps.gov/dena/)
- Denali National Park travel guide from Wikitravel
- NPS Education packet (http://www.nps.gov/dena/planyourvisit/upload/Ed Packet (revised 2008,msj).pdf)

George Parks Highway

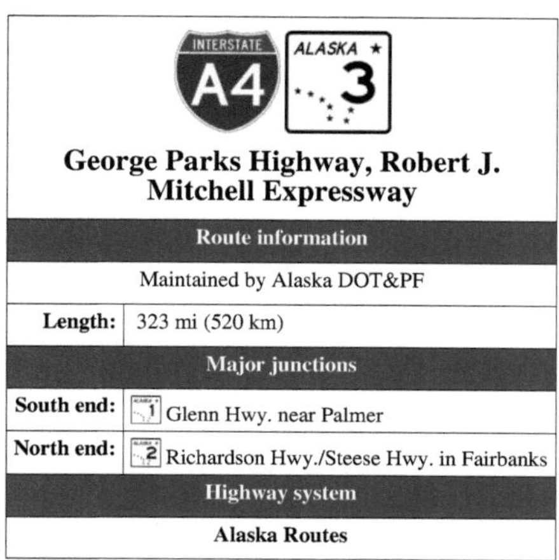

George Parks Highway, Robert J. Mitchell Expressway

Route information	
Maintained by Alaska DOT&PF	
Length:	323 mi (520 km)
Major junctions	
South end:	Glenn Hwy. near Palmer
North end:	Richardson Hwy./Steese Hwy. in Fairbanks
Highway system	
Alaska Routes	

The **George Parks Highway** (numbered **Interstate A-4** and **Alaska Route 3**), usually called simply the **Parks Highway**, runs 323 miles (520 km) from the Glenn Highway 35 miles (56 km) north of Anchorage to Fairbanks in the Alaska Interior. The highway, originally known as the Anchorage-Fairbanks Highway, was completed in 1971, and given its current name in 1975.

The Parks Highway, near Hurricane, Alaska.

The highway, which mostly parallels the Alaska Railroad, is one of the most important roads in Alaska. It is the main route between Anchorage and Fairbanks (Alaska's two largest metropolitan areas), the principal access to Denali National Park and Preserve and Denali State Park, and the main highway in the Matanuska-Susitna Valley.

It is a common misconception that the name "Parks Highway" comes from the road's proximity to the Denali state and national parks; it is in fact in honor of George Alexander Parks, governor of the Territory of Alaska from 1925 to 1933.[1] However, the aptness of the name was recognized when it was chosen.

Mileposts along the Parks Highway do not begin with 0 (zero). Instead, they begin with Mile 35 (km 56), continuing the milepost numbering of the Glenn Highway where the two highways intersect near Palmer. The 0 (zero) mile marker for the Glenn Highway is at its terminus in downtown Anchorage at the intersection of East 5th Avenue and Gambell Street. Thus mileposts along the Parks Highway reflect distance from Anchorage, which is not actually on the Parks Highway.

There are two sections of the highway that are built to freeway standards. These include an area near the highway's intersection with the Glenn Highway in Palmer and a stretch known as the Mitchell Expressway in Fairbanks leading to the highway's junction with the Richardson Highway.

Controversy

Former Alaska governor Tony Knowles criticized Sarah Palin for supporting the Knik Arm Bridge, the Gravina Island Bridge, and a road north out of Juneau instead of rebuilding the Parks Highway. However, the *Ketchikan Daily News* noted that, of the gubernatorial candidates, "Only Palin is consistent in support all of the projects".[2] [3] [4] [5] [6]

Exit list

Exits in Alaska are not numbered.

Parks Highway headed toward Fairbanks

Milepost	Street	Notes
35.0	**Traffic Is Directed Onto 1 Glenn Hwy South**	
35.0	1 Glenn Hwy North	Southbound Only
35.92	Matanuska Rd.	
37.58	N. Hyer Road	
38.66	**Palmer Section Of Freeway Ends**	
352.65	**Fairbanks Section Of Freeway Begins**	
352.65	Geist Road	
353.8_	Airport Road	
	Gap In Freeway	
354.7_	University Ave. South	
356.94	Peger Rd.	
357.95	Lathrop St.	
	Freeway Resumes	
358.0	2 Steese Hwy N.	Northbound Only
358.0	**Traffic Directed Onto 2 Richardson Hwy S.**	

Towns and places along the Parks Highway

Parks Highway milepost 238 marker, near Denali
National Park

- Wasilla, mile 42 (km 68)
- Big Lake, via Big Lake Road, mile 52 (km 84)
- Houston, mile 57 (km 92)
- Willow, mile 69 (km 111)
- Hatcher Pass, via Hatcher Pass Road, mile 71 (km 115)
- Talkeetna, via Talkeetna Spur Road, mile 99 (km 159)
- Trapper Creek, mile 115 (km 185)
- Denali State Park, miles 132–169 (km 212–272)
- Cantwell, mile 210 (km 338)
- Denali National Park entrance, mile 237 (km 382)
- Healy, mile 249 (km 400)
- Anderson and Clear Air Force Station, mile 284 (km 456)
- Nenana, mile 304 (km 490)
- Ester, mile 352 (km 566)
- Fairbanks, mile 358 (km 576)

References

[1] "Parks Highway" (http://news.google.com/newspapers?id=oD8dAAAAIBAJ&sjid=16YEAAAAIBAJ&pg=5027,1440414). *Anchorage Daily News*: p. 4. July 16, 1975. .

[2] "Palin Criticized during gubernatorial campaign for her support of Gravina Island Bridge" (http://archive.ketchikandailynews.com/ archive_detail.php?archiveFile=./pubfiles/kdn/archive/2006/October/28/Unspecified/10282006_B-02.pdf.xml). 2006-10-26. .

[3] "Palin voiced initial support for the proposed Gravina Island bridge during campaign" (http://archive.ketchikandailynews.com/ archive_detail.php?archiveFile=./pubfiles/kdn/archive/2006/September/21/LocalNews/5472.xml). 2006-09-21. .

[4] "Palin defends the bridge project, asks people to band together" (http://archive.ketchikandailynews.com/archive_detail.php?archiveFile=. /pubfiles/kdn/archive/2006/October/02/LocalNews/5692.xml). 2006-10-02. .

[5] (http://google.com/search?q=cache:nPTPaW0w62sJ:archive.ketchikandailynews.com/archive_detail.php?archiveFile=./pubfiles/kdn/ archive/2006/October/21/StateNews/6220.xml&start=0&numPer=20&keyword=supports+the+Ketchikan+bridge+project& sectionSearch=&begindate=1%2F1%2F2003&enddate=9%2F3%2F2008&authorSearch=&IncludeStories=1&pubsection=&page=& IncludePages=1&IncludeImages=1&mode=allwords) Candidate Palin Supported the Gravina Island Bridge project days before gubernatorial election.

[6] "Palin Criticized during gubernatorial campaign for her support of Gravina Island Bridge" (http://archive.ketchikandailynews.com/ archive_detail.php?archiveFile=./pubfiles/kdn/archive/2006/October/28/Unspecified/10282006_B-02.pdf.xml). 2006-10-28. .

External links

- A journey down the George Parks Highway (http://members.virtualtourist.com/m/tt/8fd07/#TL)

Tangle Lakes

The **Tangle Lakes** (Long Tangle Lake, Lower Tangle Lake, Round Tangle Lake, and Upper Tangle Lake[1]) are a 16 miles (26 km) long chain of lakes connected by streams in interior Alaska. They form the headwaters for the Delta River.

The main public access to the Lakes is from a Bureau of Land Management maintained campground and boat launch at Round Tangle Lake, about 20 miles (32 km) from Paxson on the Denali Highway. The boat launch is also the upper terminus of the Delta River Canoe Trail, a 2-3 day route to the Gulkana River and the Richardson Highway.[2] The lakes support many species of fish, including lake trout, burbot, and Arctic grayling. The area around the lakes consists mostly of tundra due to the high elevation (2864 feet (873 m)).

Round Tangle Lake with nearby peaks of the Alaska Range

The Tangle Lakes area has been the subject of extensive archaeological exploration. Prior to 1976 almost 150 sites had been discovered showing that The Tangle Lakes have been populated intermittently since the settlement of the new world. The sites are concentrated closely to the lakes in a range covering about 80 square kilometers and the attraction of the location was most likely that the windswept hills surrounding the lakes would have attracted caribou seeking to graze on the exposed lichen (not for fishing as the lakes would not have been able to support a settlement).[3]

References

[1] The Milepost 61st Edition, page 502 ISBN 978-1-892-15426-2
[2] U.S. Geological Survey Geographic Names Information System: Tangle Lakes (http://geonames.usgs.gov/pls/gnispublic/ f?p=gnispq:3:::NO::P3_FID:1410634)
[3] F. H. West. "Old World Affinities of Archaeological Complexes from Tangle Lakes (Central Alaska)". Beringia in the Cenozoic Era. Ed. V. L. Kontrimavichus. New Delhi: Amerind Publishing Co. Pvt. Ltd., 1984. 571-596.

Further reading

- Pellerin, L. (2003). Magnetotelluric data in the Delta River Mining District, near the Tangle Lakes area of south-central Alaska [Open-file Report 03-328]. Denver, CO: U.S. Department of the Interior, U.S. Geological Survey.

External links

- Fish Alaska Magazine (http://www.fishalaskamagazine.com/archives/2003/0403_tanglelakes.htm)
- Lonely Planet Alaska (http://books.google.com/books?id=b-JDesZWm5gC&pg=PA332)
- BLM Alaska (http://www.blm.gov/ak/st/en/prog/sa/delta_nwsr/lowertangle.html)
- Unofficial Guide to Adventure Travel in Alaska (http://books.google.com/books?id=j0V-9ktou64C& pg=PA136)
- Archaeology of Prehistoric Native Alaska (http://books.google.com/books?id=_0u2y_SVnmoC& pg=RA1-PA828)
- BLM Tangle Lakes Archaeological District trail brochure (http://www.blm.gov/pgdata/etc/medialib/blm/ak/ gdo/pdf_files.Par.12713.File.dat/05TLAD_trails_brochure.pdf)

Copper River (Alaska)

<table>
<tr><td colspan="2" align="center">Copper River</td></tr>
<tr><td colspan="2">
A fisherman (bottom center) dipnetting for salmon on the Copper River at Chitina in Southcentral Alaska</td></tr>
<tr><td>Origin</td><td>62°10′39″N 143°49′05″W Copper Glacier on Mount Wrangell</td></tr>
<tr><td>Mouth</td><td>60°23′19″N 144°57′39″W Copper Bay of Pacific Ocean</td></tr>
<tr><td>Basin countries</td><td>United States of America</td></tr>
<tr><td>Length</td><td>286 mi (460 km)</td></tr>
<tr><td>Source elevation</td><td>4380 ft (1340 m)</td></tr>
<tr><td>Mouth elevation</td><td>0 ft (0 m)</td></tr>
<tr><td>Avg. discharge</td><td>56000 cu ft/s (1600 m^3/s) at mouth</td></tr>
<tr><td>Basin area</td><td>24400 sq mi (63000 km^2)</td></tr>
</table>

The **Copper River** or **Ahtna River** (Ahtna Athabascan **'Atna**) is a 300-mile (480 km) river in south-central Alaska in the United States. It drains a large region of the Wrangell Mountains and Chugach Mountains into the Gulf of Alaska. It is known for its extensive delta ecosystem, as well as for its prolific runs of wild salmon, which are among the most highly prized stocks in the world. It is the tenth largest river in the United States, as ranked by average discharge volume at its mouth.[1]

Description

The Copper River rises out of the Copper Glacier, which lies on the northeast side of Mount Wrangell, in the Wrangell Mountains, within Wrangell-Saint Elias National Park. It begins by flowing almost due north in a valley that lies on the east side of Mount Sanford, and then turns west, forming the northwest edge of the Wrangell Mountains and separating them from the Mentasta Mountains to the northeast. It continues to turn southeast, through a wide marshy plain to Chitina, where it is joined from the southeast by the Chitina River. The Copper River is 287 miles (462 km) long. It drops an average of about 12 feet per mile (2.3 m/km), and drains a total of 24000 square miles (62000 km^2)—an area the size of West Virginia. The river has 13 major tributaries and runs at an average of 7 miles per hour (11 km/h). It is a mile (1.6 km) wide at the Copper River Delta, near Cordova. Downstream from its confluence with the Chitina it flows southwest, passing through a narrow glacier-lined gap in the Chugach Mountains east of Cordova Peak. There is an extensive area of sand dunes between the Copper and Bremner Rivers. Both Miles Glacier and Child's Glacier calve directly into the river. The Copper enters the Gulf of Alaska approximately 50 miles (80 km) southeast of Cordova.

The name of the river comes for the abundant copper deposits along the upper river that were used by Alaska Native population and then later by settlers from the Russian Empire and the United States. Extraction of the copper resources was rendered difficult by navigation difficulties at the river's mouth. The construction of the Copper River and Northwestern Railway from Cordova through the upper river valley in 1908-11 allowed widespread extraction of the mineral resources, in particular from the Kennecott Mine, discovered in 1898. The mine was abandoned in 1938 and is now a ghost town tourist attraction. A road runs from Cordova to the lower Copper River near Child's Glacier, following the old railroad route and ending at the reconstructed "Million Dollar Bridge" across the river. The Tok Cut-Off follows the Copper River Valley on the north side of the Chugach Mountains.

The river's famous salmon runs arise from the use of the river watershed by over 2 million salmon each year for spawning. The extensive runs result in many unique varieties. The river's commercial salmon season is short: chinook (king) salmon are available mid-May to mid-June, sockeye (red) salmon mid-May to mid-August, and coho (silver) salmon mid-August to late-September. Sport, personal use and subsistence fisheries are open for salmon from mid-May through October. The fisheries are co-managed by the Alaska Department of Fish and Game and the USDA Forest Service Federal Subsistence Board. Management data are obtained primarily by ADF&G at the Miles Lake Sonar Station and the Native Village of Eyak at the Baird Canyon/ Canyon Creek research stations.

The Copper River Delta, which extends for 700,000 acres (2,800 km^2) is the considered the largest contiguous wetlands along the Pacific coast of North America. It is used annually by 16 million shorebirds, including the world's entire population of western sandpipers. It is also home to the world's largest population of nesting trumpeter swans and is the only known nesting site for the dusky Canada goose. Over 20,000 years ago, the area now drained by the great Copper River was a massive lake, covering 2000 square miles (5200 km^2).

Miles Glacier Bridge, showing earthquake damage and temporary repair, 1984

Rafters and Child's Glacier on the lower Copper River.

Fishwheels on the Copper River

Sand dunes on the Copper River.

Copper River near Chitina, looking south from the bridge

Black Spruce taiga along the Copper River.

Wind picks up fine sediment from the riverbank and carries it over the ocean.

See also

* List of Alaska rivers

References

[1] "Largest Rivers in the United States" (http://pubs.usgs.gov/of/1987/ofr87-242/) (PDF). USGS. .

Further reading

* Brabets, T.P. (1997). *Geomorphology of the lower Copper River, Alaska* [U.S. Geological Survey Professional Paper 1581]. Washington, D.C.: U.S. Department of the Interior, U.S. Geological Survey.

External links

* Ecotrust Copper River Program (http://www.ecotrust.org/copperriver)
* Copper River salmon habitat management study / prepared for Ecotrust ; prepared by Marie E. Lowe of the Institute of Social and Economic Research (http://library.state.ak.us/asp/edocs/2007/04/ocn123246851.pdf) Hosted by Alaska State Publications Program (http://library.state.ak.us).
* Alaska Department of Fish and Game: Copper River Salmon (http://www.cf.adfg.state.ak.us/region2/crhome.php)
* Ghost Town of the Kennecott Copper Mine (http://www.alaskagold.com/copper/mcarthy/mcarthy.html)
* Packrafting Journey down the Copper River (http://www.aktrekking.com/2004/Copper/Copper1.html)
* Eyak Preservation Council (http://www.redzone.org/)
* Nature Conservancy: Copper River Delta (http://nature.org/wherewework/northamerica/states/alaska/preserves/art11190.html)
* The Copper River Watershed Project (http://www.copperriver.org/links.html)
* NVE Fisheries Research and Seasonal Employment on the Copper River (http://www.eyakfish.com)
* Cordova District Fishermen United (http://www.crsalmon.org)
* Wrangell-St. Elias National Park information (http://www.largestnationalpark.com)
* Copper River | Chitina Dipnet Fishery Escapement Charts (http://aksalmondipnetting.com/copper/index.htm)
* Copper River General Information (http://www.guidesinalaska.com/river.ak?ID=8087&name=Copper-River-Alaska)

Mount Deborah

Mount Deborah	
Elevation	12339 ft (3761 m)
Prominence	5200 ft (1585 m)
Location	
Location	Denali Borough, Alaska, USA
Range	Eastern Alaska Range
Coordinates	63°38′16″N 147°14′20″W
Topo map	USGS Healy C-1
Climbing	
First ascent	1954 by Fred Beckey, Henry Meybohm, Heinrich Harrer
Easiest route	snow/ice climb

Mount Deborah is a mountain in the U.S. state of Alaska. It is one of the major peaks of the eastern Alaska Range. Despite its low absolute elevation (compared to other major peaks in North America), it is a particularly large and steep peak in terms of its quick rise over local terrain. For example, the Northeast Face rises 7000 feet (2135m) in approximately 1.5 miles (2.4 km). This steepness, combined with difficult access, harsh weather, and classic Alaskan ice and snow features, make this a challenging peak to climb.[1]

Mount Deborah was named in 1907 by James Wickersham for his first wife, Deborah Susan (Bell) Wickersham.[1][2]

First ascent

Mount Deborah was first climbed in 1954 by Fred Beckey, Henry Meybohm, Heinrich Harrer, via the South Ridge.

Notable subsequent attempts and ascents

- 1964 *East Ridge* (attempt) by David Roberts and Don Jensen.[3]
- 1975 *West Buttress and South Ridge* FA of route, 2nd of summit by Pat Condran, Mark Hottman, Brian Okonek, Dave Pettigrew, Pat Stuart and Toby Wheeler. Summit reached May 8 after 36 days of effort.[4]
- 1977 *North Face* FA of route by Dakers Gowans and Charles Macquarrie, summit reached May 8.[5]

Today's standard route is the West Face; the Northwest Ridge is also a recommended route.

Mount Deborah is the subject of one of the classics of mountaineering literature, *Deborah: A Wilderness Narrative*, by David Roberts, which describes a failed attempt on the peak in 1964.

Deborah Avenue in College, Alaska is named for the mountain. Parallel to the street in the same subdivision is Hayes Avenue.

References

- Michael Wood and Colby Coombs, *Alaska: A Climbing Guide*, The Mountaineers, 2001.

[1] Orth, Donald J. (1967). *Dictionary of Alaska Place Names*. Washington: U.S. Government Printing Office. p. 262.

[2] Atwood, Evangeline; DeArmond, Robert N. (1977). *Who's Who in Alaskan Politics*. Portland: Binford & Mort for the Alaska Historical Commission. p. 106.

[3] Roberts, David (1991). *Deborah; and, The Mountain of my fear: the early climbs*. Seattle, WA, USA: The Mountaineers. ISBN 9780898862706.

[4] Okonek, Brian (1976). "Deborah". *American Alpine Journal 1976* (New York, NY, USA: American Alpine Club) **20** (50): 294–301.

[5] Gowans, Dakers (1978). "Deborah's North Face". *American Alpine Journal 1978* (New York, NY, USA: American Alpine Club) **21** (52): 337–343.

Further reading

Roberts, David (1991). *Deborah; and, The Mountain of my fear: the early climbs*. Seattle, WA, USA: The Mountaineers. ISBN 9780898862706.

External links

- Mount Deborah on Topozone (http://www.topozone.com/map.asp?lat=63.6378&lon=-147.2388& datum=NAD83&s=500&size=l&layer=DRG250)
- Alaskan peaks with prominence > 1500m (http://www.peaklist.org/USlists/AK5000.html)
- Mount Deborah on bivouac.com (http://bivouac.com/MtnPg.asp?MtnId=7705)

Wrangell Mountains

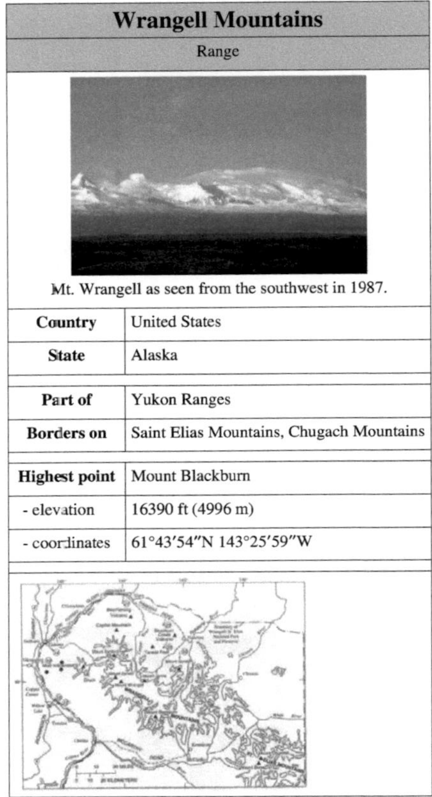

Wrangell Mountains	
Range	
Mt. Wrangell as seen from the southwest in 1987.	
Country	United States
State	Alaska
Part of	Yukon Ranges
Borders on	Saint Elias Mountains, Chugach Mountains
Highest point	Mount Blackburn
- elevation	16390 ft (4996 m)
- coordinates	61°43′54″N 143°25′59″W

The **Wrangell Mountains** are a high mountain range of eastern Alaska in the United States. Much of the range is included in Wrangell-Saint Elias National Park and Preserve. The Wrangell Mountains are almost entirely volcanic in origin, and they include the second and third highest volcanoes in the United States, Mount Blackburn and Mount Sanford. The range takes its name from Mount Wrangell, which is one of the largest andesite shield volcanoes in the world, and also the only presently active volcano in the range. The Wrangell Mountains comprise most of the Wrangell Volcanic Field, which also extends into the neighboring Saint Elias Mountains and the Yukon Territory in Canada.

The Wrangell Mountains are just to the northwest of the Saint Elias Mountains and northeast of the Chugach Mountains, which are along the coast of the Gulf of Alaska. These ranges have the combined effect of blocking the inland areas from warmer moist air over the Pacific Ocean. The inland areas to the north of the Wrangell Mountains are therefore among the coldest areas of North America during the winter.

Major peaks

Mount Sanford

Hikers on a pass between Mt. Sanford and Mt. Drum

The Wrangell Mountains include 12 of the 40+ Alaskan peaks over 13,000 ft (see fourteeners and thirteeners):

- Mount Blackburn, 16390 feet (4996 m), and East Summit, 16286 ft (4964 m)
- Mount Sanford, 16237 feet (4949 m), and South Peak, 13654 ft (4162 m)
- Mount Wrangell, 14163 feet (4317 m), and West Summit, 14013 ft (4271 m)
- Atna Peaks, 13860 ft (4225 m)
- Regal Mountain, 13845 ft (4220 m)
- Mount Jarvis, 13421 feet (4091 m), and North Peak, 13025 ft (3970 m)
- Parka Peak, 13280 ft (4048 m)
- Mount Zanetti, 13009 ft (3965 m)

Other prominent mountains include:

- Mount Drum, 12010 ft (3661 m)

Name origin and references in popular culture

Mountains named after explorer, president of Russian-American Company, admiral Ferdinand von Wrangel. American folk singer John Denver wrote a song, "Wrangell Mountain Song", in reference to the range.

References

- Richter, Donald H.; Danny S. Rosenkrans and Margaret J. Steigerwald (1995). *Guide to the Volcanoes of the Western Wrangell Mountains, Alaska* [1]. USGS Bulletin 2072.
- Winkler, Gary R. (2000). *A Geologic Guide to Wrangell—Saint Elias National Park and Preserve, Alaska: A Tectonic Collage of Northbound Terranes* [2]. USGS Professional Paper 1616. ISBN 0-607-92676-7.
- Richter, Donald H.; Cindi C. Preller, Keith A. Labay, and Nora B. Shew (2006). *Geologic Map of the Wrangell-Saint Elias National Park and Preserve, Alaska* [3]. USGS Scientific Investigations Map 2877.
- Wood, Charles A.; Jürgen Kienle, eds. (1990). *Volcanoes of North America*. Cambridge University Press. ISBN 0-521-43811-X.

External links

- Wrangell-St. Elias National Park & Preserve [4]
- Wrangell Mountains Center [5]

References

[1] http://www.nps.gov/wrst/naturescience/guide-to-the-wrangell-volcanoes.htm

[2] http://pubs.usgs.gov/pp/p1616/

[3] http://pubs.usgs.gov/sim/2006/2877/

[4] http://www.nps.gov/wrst/

[5] http://www.wrangells.org/

Chugach Mountains

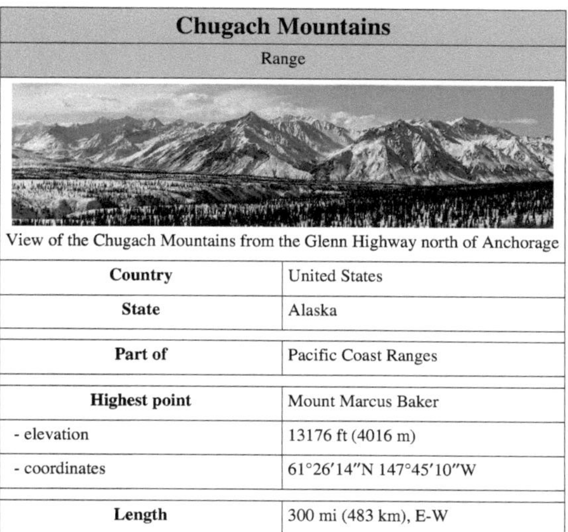

Chugach Mountains	
Range	
View of the Chugach Mountains from the Glenn Highway north of Anchorage	
Country	United States
State	Alaska
Part of	Pacific Coast Ranges
Highest point	Mount Marcus Baker
- elevation	13176 ft (4016 m)
- coordinates	61°26′14″N 147°45′10″W
Length	300 mi (483 km), E-W

The **Chugach Mountains** of southern Alaska are the northernmost of the several mountain ranges that make up the Pacific Coast Ranges of the western edge of North America. The range is about 500 km (300 mi) long, running generally east-west. Its highest point is Mount Marcus Baker, at 13176 feet (4016 m), but most of its summits are not especially high. Even so its position along the Gulf of Alaska ensures more snowfall in the Chugach than anywhere else in the world; an annual average of over 1500 cm (600 in).

Alpine Lakes in the Chugach Mountains

The mountains are protected in the Chugach State Park and the Chugach National Forest. Near to Anchorage, they are a popular destination for outdoor activities. The World Extreme Skiing Championship is held annually in the Chugach near Valdez.

The Richardson Highway, Seward Highway, and the Glenn Highway run through the Chugach Mountains. The tunnel from Portage on the Turnagain Arm of Cook Inlet to Whittier on Passage Canal also provides railroad and automobile access underneath Maynard Mountain to the Prince William Sound.

The name "Chugach" is from the Eskimo tribal name *Chugachmiut* recorded by the Russians and written by them "Chugatz" and "Tchougatskoi"; in 1898 U.S. Army Captain W. R. Abercrombie spelled the name "Chugatch" and applied it to the mountains.[1]

A peak in the Chugach Mountains

Mountains

- Mount Marcus Baker 13176 ft (4016 m)

- Mount Thor 12251 ft (3734 m)
- Mount Steller 10617 feet (3236 m)
- Mount Michelson 8701 feet (2652 m)
- Mount Palmer 6940 feet (2115 m)
- Flattop Mountain 3510 feet (1070 m)
- Eagle Peak 6955 feet (2120 m)
- Polar Bear Peak 6614 feet (2016 m)
- Ptarmigan Peak (Alaska)
- Wolverine Peak
- O'Malley Peak

See also

- Matanuska Formation

References

[1] "Chugach Mountains" (http://geonames.usgs.gov/pls/gnispublic/f?p=gnispq:3:::NO::P3_FID:1412401). Geographic Names Information System, U.S. Geological Survey. . Retrieved 2010-03-01.

Alaska Range

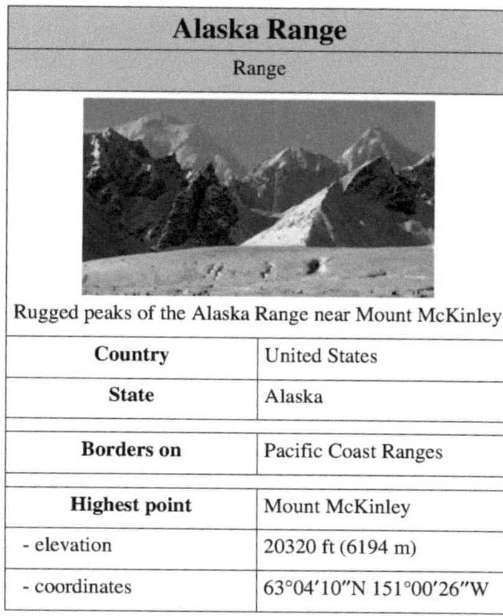

Alaska Range	
Range	
Rugged peaks of the Alaska Range near Mount McKinley	
Country	United States
State	Alaska
Borders on	Pacific Coast Ranges
Highest point	Mount McKinley
- elevation	20320 ft (6194 m)
- coordinates	63°04′10″N 151°00′26″W

The **Alaska Range** is a relatively narrow, 650-km-long (400 mi) mountain range in the southcentral region of the U.S. state of Alaska, from Lake Clark at its southwest end[1] to the White River in Canada's Yukon Territory in the southeast. The highest mountain in North America, Denali, (or Mount McKinley), is in the Alaska Range.

Description and history

The range forms a generally east-west arc with its northernmost part in the center, and from there trending southwest towards the Alaska Peninsula and the Aleutians, and trending southeast into the Pacific Coast Ranges. The mountains act as a high barrier to the flow of moist air from the Gulf of Alaska northwards, and thus has some of the harshest weather in the world. The heavy snowfall also contributes to a number of large glaciers, including the Canwell, Castner, Black Rapids, Susitna, Yanert, Muldrow, Eldridge, Ruth, Tokositna, and Kahiltna Glaciers. Four major rivers cross the Range, including the Delta River, and Nenana River in the center of the range and the Nabesna and Chisana Rivers to the east.

View from Denali State Park

Alaska Range Glacier

The range is part of the Pacific Ring of Fire, and the Denali Fault that runs along the southern edge of the range is responsible for a number of earthquakes. Mount Spurr is a stratovolcano located in the northeastern end of the Aleutian Volcanic Arc of Alaska, USA which has two vents, the summit and nearby Crater Peak.

Parts of the range are protected within Wrangell-St. Elias National Park and Preserve, Denali National Park and Preserve, and Lake Clark National Park and Preserve. The George Parks Highway from Anchorage to Fairbanks, the Richardson Highway from Valdez to Fairbanks, and the Tok Cut-Off from Gulkana Junction to Tok, Alaska pass through low parts of the range. The Alaska Pipeline parallels the Richardson Highway.

Naming history

The name "Alaskan Range" appears to have been first applied to these mountains in 1869 by naturalist W. H. Dall. The name eventually became "Alaska Range" through local use. In 1849 Constantin Grewingk applied the name "Tschigmit" to this mountain range. A map made by the General Land Office in 1869 calls the southwestern part of the Alaska Range the "Chigmit Mountains" and the northeastern part the "Beaver Mountains".[2] However the Chigmit Mountains are now considered part of the Aleutian Range.

Major peaks

- Mount McKinley (6,194 m/20,320 ft)
- Mount Foraker (5,304 m/17,400 ft)
- Mount Hunter (4,442 m/14,573 ft)
- Mount Hayes (4,216 m/13,832 ft)
- Mount Silverthrone (4,029 m/13,218 ft)
- Mount Deborah (3,761 m/12,339 ft)
- Mount Huntington (3,730 m/12,240 ft)
- Mount Russell (3,557 m/11,670 ft)

Mount McKinley, on a rare clear day

Subranges (from west to east)

- Neacola Mountains[1]
- Revelation Mountains
- Teocalli Mountains
- Kichatna Mountains
- Central Alaska Range/Denali Massif
- Eastern Alaska Range/Hayes Range
- Delta Mountains
- Mentasta Mountains
- Nutzotin Mountains

Alaska Range Mountain Peaks

Documented wilderness traverses of Alaska Range

- Mentasta Lake to Kitchatna Mountains (1981): Scott Woolums, George Beilstein, Steve Eck, and Larry Coxen by skis: first traverse. 375 miles (604 km) in 45 days.[3]
- Canada to Lake Clark (1996): Roman Dial, Carl Tobin, and Paul Adkins by mountain bike and packraft: first full length traverse. 775 miles (1247 km) in 42 days.[4]
- Tok to Lake Clark (1996): Kevin Armstrong, Doug Woody, and Jeff Ottmers by snowshoe, foot, and packraft: first foot traverse. 620 miles (1000 km) in 90 days.[5]

The Denali Highway passes through the Alaska Range and offers travelers a close up-look at some of the lower peaks

References

Gulkana Glacier flows from the ice fields of the Alaska Range

[1] Sources differ as to the exact delineation of the Alaska Range. The Board on Geographic Names (http://geonames.usgs.gov/domestic/index.html) entry is inconsistent; part of it designates Iliamna Lake as the southwestern end, and part of the entry has the range ending at the Telaquana and Neacola Rivers. Other sources identify Lake Clark, in between those two, as the endpoint. This also means that the status of the Neacola Mountains is unclear: it is usually identified as the northernmost subrange of the Aleutian Range, but it could also be considered the southernmost part of the Alaska Range.

[2] Name history from the Board on Geographic Names (http://geonames.usgs.gov/domestic/index.html) entry for the Alaska Range.

[3] American Alpine Journal (1982), Vol. 24. Pages 137-138

[4] "A Wild Ride," National Geographic Magazine (1997), Vol. 191. Pages 118-131

[5] American Alpine Journal (1997), Vol. 39. Pages 169-170

Further reading

- Churkin, M., Jr., and C. Carter. (1996). *Stratigraphy, structure, and graptolites of an Ordovician and Silurian sequence in the Terra Cotta Mountains, Alaska Range, Alaska* [U.S. Geological Survey Professional Paper 1555]. Washington, D.C.: U.S. Department of the Interior, U.S. Geological Survey.

Alaska

<table>
<tr><td colspan="2" align="center">State of Alaska</td></tr>
<tr><td align="center">
Flag</td><td align="center">
Seal</td></tr>
<tr><td colspan="2" align="center">Nickname(s): The Last Frontier</td></tr>
<tr><td colspan="2" align="center">Motto(s): North to the Future</td></tr>
<tr><td colspan="2" align="center">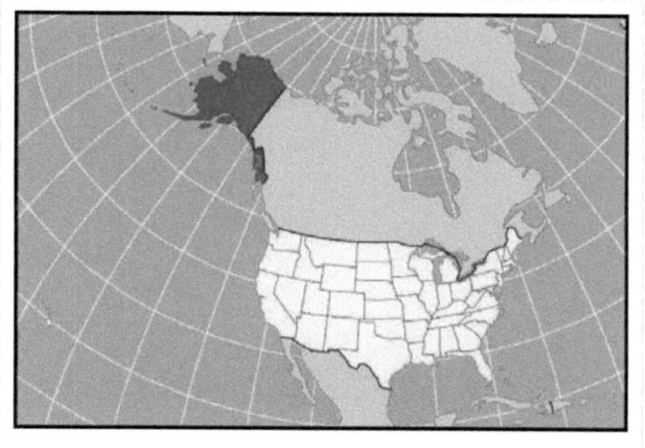</td></tr>
</table>

Official language(s)	None[1] [2]
Spoken language(s)	English 89.7%, Native North American 5.2%, Spanish 2.9%
Demonym	Alaskan
Capital	Juneau
Largest city	Anchorage
Area	Ranked 1st in the U.S.
- Total	663,268 sq mi (1,717,854 km^2)
- Width	2,261 miles (3,639 km)
- Length	1,420 miles (2,285 km)
- % water	13.77
- Latitude	51°20'N to 71°50'N
- Longitude	130°W to 172°E
Population	Ranked 47th in the U.S.
- Total	710,231 (2010)[3] 626,932 (2000)

- Density	1.07/sq mi (0.4/km^2) Ranked 50th in the U.S.
- Median income	US$64,333 (4th)
Elevation	
- Highest point	Mount McKinley (Denali)[4] 20,320 ft (6194 m)
- Mean	1900 ft (580 m)
- Lowest point	Ocean[4] sea level
Admission to Union	January 3, 1959 (49th)
Governor	Sean Parnell (R)
Lieutenant Governor	Mead Treadwell (R)
Legislature	Alaska Legislature
- Upper house	Senate
- Lower house	House of Representatives
U.S. Senators	Lisa Murkowski (R) Mark Begich (D)
U.S. House delegation	Don Young (R) (at-large) (list)
Time zones	
- east of 169° 30'	Alaska: UTC-9/DST-8
- west of 169° 30'	Aleutian: UTC-10/DST-9
Abbreviations	AK US-AK
Website	[www.alaska.gov www.alaska.gov]

Alaska ◄) [i]/əˈlæskə/ is the largest state in the United States by area. It is situated in the northwest extremity of the North American continent, with Canada to the east, the Arctic Ocean to the north, and the Pacific Ocean to the west and south, with Russia further west across the Bering Strait. Approximately half of Alaska's 710,231 residents (as per the 2010 United States Census) live within the Anchorage metropolitan area. Alaska is the least densely populated state of the U.S.[5]

Alaska was purchased from Russia on March 30, 1867, for $7.2 million ($113 million in today's dollars) at approximately two cents per acre ($4.74/km²). The land went through several administrative changes before becoming an organized (or incorporated) territory on May 11, 1912, and the 49th state of the U.S. on January 3, 1959.

The name "Alaska" (Аляска) was already introduced in the Russian colonial period, when it was used only for the peninsula and is derived from the Aleut *alaxsxaq*, meaning "the mainland" or more literally, "the object towards which the action of the sea is directed".[6] It is also known as Alyeska, the "great land", an Aleut word derived from the same root.

Geography

Alaska has a longer coastline than all the other U.S. states combined.[7] It is the only non-contiguous U.S. state on continental North America; about 500 miles (800 km) of British Columbia (Canada) separate Alaska from Washington state. Alaska is thus an exclave of the United States. It is technically part of the continental U.S., but is often not included in colloquial use; Alaska is not part of the contiguous U.S., often called "the Lower 48".[8] The capital city, Juneau, is situated on the mainland of the North American continent, but is not connected by road to the rest of the North American highway system.

The state is bordered by the Yukon Territory and British Columbia in Canada, to the east, the Gulf of Alaska and the Pacific Ocean to the south, the Bering Sea, Bering Strait, and Chukchi Sea to the west and the Arctic Ocean to the north. Alaska's territorial waters touch Russia's territorial waters in the Bering Strait, as the Russian Big Diomede Island and Alaskan Little Diomede Island are only 3 miles (4.8 km) apart. With the extension of the Aleutian Islands into the eastern hemisphere, it is technically both the westernmost and easternmost state in the United States, as well as also being the northernmost.

Alaska is the largest state in the United States in land area at 586412 square miles (km^2), over twice the size of Texas, the next largest state. Alaska is larger than all but 18 sovereign countries. Counting territorial waters, Alaska is larger than the combined area of the next three largest states: Texas, California, and Montana. It is also larger than the combined area of the 22 smallest U.S. states.

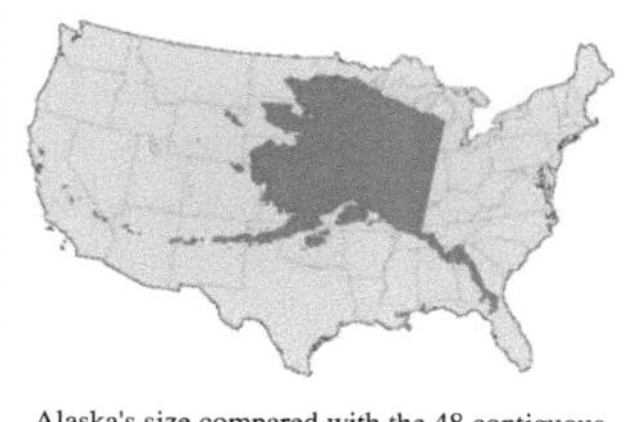

Alaska's size compared with the 48 contiguous states.

Regions

There are no officially defined borders demarcating the various regions of Alaska, but there are six generally accepted regions:

South Central

The most populous region of Alaska, containing Anchorage, the Matanuska-Susitna Valley and the Kenai Peninsula. Rural, mostly unpopulated areas south of the Alaska Range and west of the Wrangell Mountains also fall within the definition of Southcentral, as well as the Prince William Sound area and the communities of Cordova and Valdez.

Southeast

Also known as the Panhandle, this is the region of Alaska closest to the rest of the United States, and hence was where most initial non-Native settlement occurred following the Alaska Purchase. It contains the state capital, Juneau, the former capital, Sitka, and the large town of Ketchikan. The road systems leading from these cities are strictly local; no roads connect these communities to each other or any other communities apart from their own suburbs. The region is dominated by the Alexander Archipelago as well as the Tongass National Forest, the largest national forest in the United States.

Interior

The largest region of Alaska, much of it uninhabited wilderness. Fairbanks is the only community of any significant size. Small towns and Native villages are scattered throughout, mostly along the highway and river systems. Denali National Park and Preserve is located here, home to Mount McKinley (also widely known by its local name of Denali), the highest point in North America.

Mount McKinley is both the highest peak in Alaska and in all of North America.

Southwest

A sparsely inhabited region stretching some 500 miles (800 km) inland from the Bering Sea. Most of the population lives along the coast. Kodiak Island is also located in Southwest. The massive Yukon–Kuskokwim Delta, one of the largest river deltas in the world, is here. Portions of the Alaska Peninsula are considered part of Southwest, with the remaining portions included with the Aleutian Islands (see below).

North Slope

The North Slope is mostly tundra peppered with small villages. The area is known for its massive reserves of crude oil, and contains both the National Petroleum Reserve–Alaska and the Prudhoe Bay Oil Field.[9] Barrow, the northernmost city in the United States, is located here. The Northwest Arctic area, anchored by Kotzebue and also containing the Kobuk River valley, is often regarded as being part of this region. However, the respective Inupiat of the North Slope and of the Northwest Arctic seldom think of themselves as one.

Grizzly bear fishing for salmon at Brooks Falls.

Aleutian Islands

More than 300 small, volcanic islands make up this chain, which stretches over 1200 miles (1900 km) into the Pacific Ocean. The International Date Line was drawn west of 180° to keep the whole state, and thus the entire North American continent, within the same legal day. However, because some of these islands fall in the Eastern Hemisphere, this makes Alaska the northernmost, easternmost and westernmost state in the union, with the southernmost state being Hawaii. Two of the islands, Attu and Kiska, were occupied by Japanese forces during World War II.

Natural features

With its myriad islands, Alaska has nearly 34000 miles (54720 km) of tidal shoreline. The Aleutian Islands chain extends west from the southern tip of the Alaska Peninsula. Many active volcanoes are found in the Aleutians and in coastal regions. Unimak Island, for example, is home to Mount Shishaldin, which is an occasionally smoldering volcano that rises to 10000 feet (3048 m) above the North Pacific. It is the most perfect volcanic cone on Earth, even more symmetrical than Japan's Mount Fuji. The chain of volcanoes extends to Mount Spurr, west of Anchorage on the mainland. Geologists have identified Alaska as part of Wrangellia, a large region consisting of multiple states and Canadian provinces in the Pacific Northwest which is actively undergoing continent building.

Augustine Volcano erupting on January 12, 2006.

One of the world's largest tides occurs in Turnagain Arm, just south of Anchorage – tidal differences can be more than 35 feet (10.7 m). (Many sources say Turnagain has the second-greatest tides in North America, but several areas in Canada have larger tides.)[10]

Alaska has more than three million lakes.[11] [12] Marshlands and wetland permafrost cover 188320 square miles (km^2) (mostly in northern, western and southwest flatlands). Glacier ice covers some 16000 square miles (41440 km^2) of land and 1200 square miles (3110 km^2) of tidal zone. The Bering Glacier complex near the southeastern border with Yukon covers 2250 square miles (5827 km^2) alone. With over 100,000, Alaska has half of the world's glaciers.

Land ownership

According to an October 1998 report by the United States Bureau of Land Management, approximately 65% of Alaska is owned and managed by the U.S. federal government as public lands, including a multitude of national forests, national parks, and national wildlife refuges. Of these, the Bureau of Land Management manages 87 million acres (35 million hectares), or 23.8% of the state. The Arctic National Wildlife Refuge is managed by the United States Fish and Wildlife Service. It is the world's largest wildlife refuge, comprising 16 million acres (6.5 million hectares).

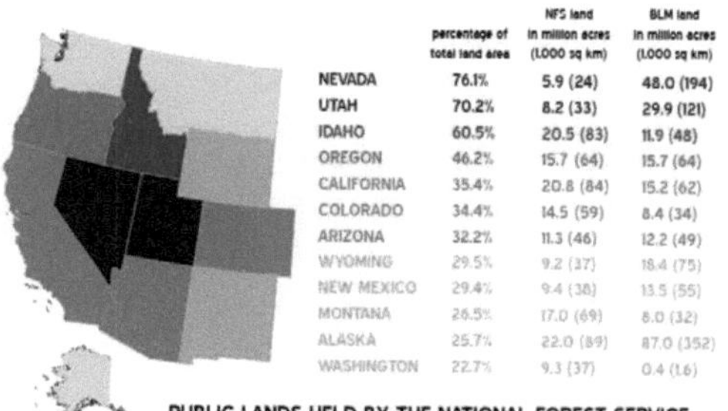

	percentage of total land area	NFS land in million acres (1,000 sq km)	BLM land in million acres (1,000 sq km)
NEVADA	76.1%	5.9 (24)	48.0 (194)
UTAH	70.2%	8.2 (33)	29.9 (121)
IDAHO	60.5%	20.5 (83)	11.9 (48)
OREGON	46.2%	15.7 (64)	15.7 (64)
CALIFORNIA	35.4%	20.8 (84)	15.2 (62)
COLORADO	34.4%	14.5 (59)	8.4 (34)
ARIZONA	32.2%	11.3 (46)	12.2 (49)
WYOMING	29.5%	9.2 (37)	18.4 (75)
NEW MEXICO	29.4%	9.4 (38)	13.5 (55)
MONTANA	26.5%	17.0 (69)	8.0 (32)
ALASKA	25.7%	22.0 (89)	87.0 (352)
WASHINGTON	22.7%	9.3 (37)	0.4 (1.6)

PUBLIC LANDS HELD BY THE NATIONAL FOREST SERVICE AND THE BUREAU OF LAND MANAGEMENT

Source: http://www.wildlandfire.com/docs/2007/western-states-data-public-land.htm

Alaska has more public land owned by the federal government than any other state.[13]

Of the remaining land area, the state of Alaska owns 101 million acres (41 million hectares); its entitlement under the Alaska Statehood Act. A portion of that acreage is occasionally ceded to organized boroughs, under the statutory provisions pertaining to newly-formed boroughs. Smaller portions are set aside for rural subdivisions and other homesteading-related opportunities, though these are infrequently popular due to the often remote and roadless locations. The University of Alaska, as a land grant university, also owns substantial acreage which it manages independently.

Another 44 million acres (18 million hectares) are owned by 12 regional, and scores of local, Native corporations created under the Alaska Native Claims Settlement Act. Regional Native corporation Doyon, Limited often promotes itself as the largest private landowner in Alaska in advertisements and other communications. Provisions of ANCSA allowing the corporations' land holdings to be sold on the open market starting in 1991 were repealed before they could take effect. Effectively, the corporations hold title (including subsurface title in many cases, a privilege denied to individual Alaskans) but cannot sell the land. Individual Native allotments can be and are sold on the open market, however.

Various private interests own the remaining land, totaling about one percent of the state. Alaska is, by a large margin, the state with the smallest percentage of private land ownership when Native corporation holdings are excluded.

Climate

The climate in Juneau and the southeast panhandle is a mid-latitude oceanic climate (Köppen climate classification *Cfb*) in the southern sections and a subarctic oceanic climate (Köppen *Cfc*) in the northern parts. On an annual basis, the panhandle is both the wettest and warmest part of Alaska with milder temperatures in the winter and high precipitation throughout the year. Juneau averages over 50 inches (1270 mm) of precipitation a year, while other areas receive over 275 inches (6990 mm).[14] This is also the only region in Alaska in which the average daytime high temperature is above freezing during the winter months.

The climate of Anchorage and south central Alaska is mild by Alaskan standards due to the region's proximity to the seacoast. While the area gets less rain than southeast Alaska, it gets more snow, and days tend to be clearer. On average, Anchorage receives 16 inches (406 mm) of precipitation a year, with around 75 inches (191 cm) of snow, although there are areas in the south central which receive far more snow. It is a subarctic climate (Köppen *Dfc*) due to its brief, cool summers.

The climate of Western Alaska is determined in large part by the Bering Sea and the Gulf of Alaska. It is a subarctic oceanic climate in the southwest and a continental subarctic climate farther north. The temperature is somewhat moderate considering how far north the area is. This region has a tremendous amount of variety in precipitation. An area stretching from the northern side of the Seward Peninsula to the Kobuk River valley is technically a desert, with portions receiving less than 10 inches (254 mm) of precipitation annually. On the other extreme, some locations between Dillingham and Bethel average around 100 inches (2540 mm) of precipitation.[14]

Barrow, known colloquially for many years by the nickname "Top of the World," is the northernmost city in the United States.

The climate of the interior of Alaska is subarctic. Some of the highest and lowest temperatures in Alaska occur around the area near Fairbanks. The summers may have temperatures reaching into the 90s°F (the low to mid 30s °C), while in the winter, the temperature can fall below . Precipitation is sparse in the Interior, often less than 10 inches (254 mm) a year, but what precipitation falls in the winter tends to stay the entire winter.

The highest and lowest recorded temperatures in Alaska are both in the Interior. The highest is 100 °F (37.8 °C) in Fort Yukon (which is just 8 miles or 13 kilometers inside the arctic circle) on June 27, 1915,[15] [16] making Alaska tied with Hawaii as the state with the lowest high temperature in the United States.[17] [18] The lowest official Alaska temperature is in Prospect Creek on January 23, 1971,[15] [16] one degree above the lowest temperature recorded in continental North America (in Snag, Yukon, Canada).[19]

The climate in the extreme north of Alaska is Arctic (Köppen *ET*) with long, very cold winters and short, cool summers. Even in July, the average low temperature in Barrow is 34 °F (1.1 °C).[20] Precipitation is light in this part of Alaska, with many places averaging less than 10 inches (254 mm) per year, mostly as snow which stays on the ground almost the entire year.

History

Alaska natives

A modern Alutiiq dancer in traditional festival garb

Numerous indigenous peoples occupied Alaska for thousands of years before the arrival of European peoples to the area. The Tlingit people developed a matriarchal society in what is today Southeast Alaska, along with parts of British Columbia and the Yukon. Also in Southeast were the Haida, now well known for their unique arts, and the Tsimshian people, whose population were decimated by a smallpox epidemic in the 1860s. The Aleutian Islands are still home to the Aleut people's seafaring society, although they were among the first native Alaskans to be exploited by Russians. Western and Southwestern Alaska are home to the Yup'ik, while their cousins the Alutiiq lived in what is now Southcentral Alaska. The Gwich'in people of the northern Interior region are primarily known today for their dependence on the caribou within the much-contested Arctic National Wildlife Refuge. The North Slope and Little Diomede Island are occupied by the widespread Inuit people.

Colonization

Some researchers believe that the first Russian settlement in Alaska was established in 17th century.[21] According to this hypothesis, in 1648 several koches of Semyon Dezhnyov's expedition were thrown to Alaska by storm and founded this settlement. This hypothesis is based on the message of Chukchi geographer Nikolai Daurkin who had visited Alaska in 1764–1765 and reported about village on the Kheuveren river, populated by "bearded men" who "pray to the icons". Some modern researchers associate Kheuveren with Koyuk River.[22]

It is usually assumed that the first European boat to reach Alaska was «St. Gabriel» under the authority of the surveyor M. S. Gvozdev and assistant navigator I. Fyodorov on August 21, 1732 during expedition of Siberian cossak A. F. Shestakov adb Belorussian explorer D. I. Pavlutsky (1729—1735)[23]

The Russian settlement of St. Paul's Harbor, Kodiak Island, 1814.

Another European contact with Alaska occurred in 1741, when Vitus Bering led an expedition for the Russian Navy aboard the *St. Peter*. After his crew returned to Russia with sea otter pelts judged to be the finest fur in the world, small associations of fur traders began to sail from the shores of Siberia towards the Aleutian islands. The first permanent European settlement was founded in 1784. Between 1774 and 1800 Spain sent several expeditions to Alaska in order to assert its claim over the Pacific Northwest. In 1789 a Spanish settlement and fort were built in Nootka Sound. These expeditions gave names to places such as Valdez, Bucareli Sound, and Cordova. Later, the Russian-American Company carried out an expanded colonization program during the early-to-mid-19th century.

Sitka, renamed New Archangel from 1804 to 1867, on Baranof Island in the Alexander Archipelago in what is now Southeast Alaska, became the capital of Russian America and remained the capital after the colony was transferred to the United States. The Russians never fully colonized Alaska, and the colony was never very profitable.

William H. Seward, the United States Secretary of State, negotiated the Alaska Purchase (also known as Seward's Folly) with the Russians in 1867 for $7.2 million. Alaska was loosely governed by the military initially, and was administered as a district starting in 1884, with a governor appointed by the president of the United States, as well as a district court headquartered in Sitka.

For most of Alaska's first decade under the American flag, Sitka was the only community inhabited by American settlers. They organized a "provisional city government," which was Alaska's first city government, but not in a legal sense. Legislation allowing Alaskan communities to legally incorporate as cities did not come about until 1900, and home rule for cities was extremely limited or unavailable until statehood took effect.

U.S. Territory

Starting in the 1890s and stretching in some places to the early 1910s, gold rushes in Alaska and the nearby Yukon Territory brought

Miners and prospectors climb the Chilkoot Trail during the Klondike Gold Rush.

thousands of miners and settlers to Alaska. Alaska was officially incorporated as an organized territory in 1912. Alaska's capital, which had been in Sitka until the 1900 legislation mandated its transfer to Juneau (the actual move took place in 1906, after initial questions arose), begun to take shape with the construction of the Alaska Governor's Mansion that same year.

U.S. troops negotiate snow and ice during the Battle of Attu in May 1943.

During World War II, the Aleutian Islands Campaign focused on the three outer Aleutian Islands – Attu, Agattu and Kiska[24] – that were invaded by Japanese troops and occupied between June 1942 and August 1943. Unalaska/Dutch Harbor became a significant base for the U.S. Army Air Corps and Navy submariners.

The U.S. Lend-Lease program involved the flying of American warplanes through Canada to Fairbanks and thence Nome; Soviet pilots took possession of these aircraft, ferrying them to fight the German invasion of the Soviet Union. The construction of military bases contributed to the population growth of some Alaskan cities.

Statehood

Statehood for Alaska was an important cause of James Wickersham early in his tenure as a congressional delegate. Decades later, the statehood movement gained its first real momentum following a territorial referendum in 1946. The Alaska Statehood Committee and Alaska's Constitutional Convention would soon follow. Statehood supporters also found themselves fighting major battles against political foes, mostly in the U.S. Congress but also within Alaska. Statehood was approved by Congress on July 7, 1958. Alaska was officially proclaimed a state on January 3, 1959.

On April 27, 1964, the massive "Good Friday Earthquake" killed 133 people and destroyed several villages and portions of large coastal communities, mainly by the resultant tsunamis and landslides. It was the third most powerful earthquake in the recorded history of the world, with a moment magnitude of 9.2. It was over one thousand times more powerful than the 1989 San Francisco earthquake. The time of day (5:36 pm), time of year and location of the epicenter were all cited as factors in potentially sparing thousands of lives, particularly in Anchorage.

The 1968 discovery of oil at Prudhoe Bay and the 1977 completion of the Trans-Alaska Pipeline led to an oil boom. Royalty revenues from oil have funded large state budgets from 1980 onward. That same year, not coincidentally, Alaska repealed its state income tax. In 1939, the *Exxon Valdez* hit a reef in the Prince William Sound, spilling over 11000000 US gallons (42000 m^3) of crude oil over 1,100 miles (1,600 km) of coastline. Today, the battle between philosophies of development and conservation is seen in the contentious debate over oil drilling in the Arctic

National Wildlife Refuge.

Demographics

Historical populations			
Census	Pop.		%±
1880	33426		—
1890	32052		−4.1%
1900	63592		98.4%
1910	64356		1.2%
1920	55036		−14.5%
1930	59278		7.7%
1940	72524		22.3%
1950	128643		77.4%
1960	226167		75.8%
1970	300382		32.8%
1980	401851		33.8%
1990	550043		36.9%
2000	626932		14.0%
2010	710231		13.3%
1930 and 1940 censuses taken in preceding autumn Sources: 1910–2010[25]			

The United States Census Bureau, as of July 1, 2008, estimated Alaska's population at 686,293,[3] which represents an increase of 59,361, or 9.5%, since the last census in 2000.[26] This includes a natural increase since the last census of 60,994 people (that is 86,062 births minus 25,068 deaths) and a decrease due to net migration of 5,469 people out of the state.[26] Immigration from outside the U.S. resulted in a net increase of 4,418 people, and migration within the country produced a net loss of 9,887 people.[26]

In 2000 Alaska ranked the 48th state by population, ahead of Vermont and Wyoming (and Washington D.C.).[27] Alaska is the least densely populated state, and one of the most sparsely populated areas in the world, at 1.0 person per square mile (0.42/km²), with the next state, Wyoming, at 5.1 per square mile (1.97/km²). Alaska is the largest U.S. state by area, and the sixth wealthiest (per capita income). As of January 2010, the state's unemployment rate is 8.5%.[28]

Race and ancestry

According to the 2010 U.S. Census, Alaska had a population of 710,231. In terms of race and ethnicity, the state was 66.7% White (64.7% Non-Hispanic White Alone), 14.8% American Indian and Alaska Native, 5.4% Asian, 3.3% Black or African American, 1.0% Native Hawaiian and Other Pacific Islander, 1.6% from Some Other Race, and 7.3% from Two or More Races. Hispanics or Latinos of any race made up 5.5% of the population.[29]

Languages

According to the 2005–2007 American Community Survey, 84.7% of people over the age of five speak only English at home. About 3.5% speak Spanish at home. About 2.2% speak another Indo-European language at home and about 4.3% speak an Asian language at home. And about 5.3% speak other languages at home.[30]

A total of 5.2% of Alaskans speak one of the state's 22 indigenous languages, known locally as "native languages". These languages belong to two major language families: Eskimo–Aleut and Na-Dene. As the homeland of these two major language families of North America, Alaska has been described as the crossroads of the continent, providing evidence for the recent settlement of North America by way of the Bering land bridge.

Religion

Alaska has been identified, along with Pacific Northwest states Washington and Oregon, as being the least religious in the U.S.[31] [32] According to statistics collected by the Association of Religion Data Archives, about 39% of Alaska residents were members of religious congregations. Evangelical Protestants had 78,070 members, Roman Catholics had 54,359, and mainline Protestants had 37,156.[33] After Catholicism, the largest single denominations are The Church of Jesus Christ of Latter-day Saints with 30,169,[34] and Southern Baptists with 22,959. The large Eastern Orthodox (with 49 parishes and up to 50,000 followers)[35] population is a result of early Russian colonization and missionary work among Alaska Natives.[36]

St. Michael's Russian Orthodox Cathedral in Sitka.

In 1795, the First Russian Orthodox Church was established in Kodiak. Intermarriage with Alaskan Natives helped the Russian immigrants integrate into society. As a result, an increasing number of Russian Orthodox churches[37] gradually became established within Alaska. Alaska also has the largest Quaker population (by percentage) of any state.[38] In 2009 there were 6,000 Jews in Alaska (for whom observance of the mitzvah may pose special problems).[39] Estimates for the number of Alaskan Muslims range from 2,000[40] [41] to 5,000.[42] In 2010, the local Muslim community broke ground on the first mosque in the state.[43] Alaskan Hindus often share venues and celebrations with members of other religious communities including Sikhs and Jains.[44] [45] [46]

Economy

The 2007 gross state product was $44.9 billion, 45th in the nation. Its per capita personal income for 2007 was $40,042, ranking 15th in the nation. The oil and gas industry dominates the Alaskan economy, with more than 80% of the state's revenues derived from petroleum extraction. Alaska's main export product (excluding oil and natural gas) is seafood, primarily salmon, cod, Pollock and crab.

Oilfield facilities at Prudhoe Bay.

Agriculture represents only a fraction of the Alaskan economy. Agricultural production is primarily for consumption within the state and includes nursery stock, dairy products, vegetables, and livestock. Manufacturing is limited, with most foodstuffs and general goods imported from elsewhere.

Employment is primarily in government and industries such as natural resource extraction, shipping, and transportation. Military bases are a significant component of the economy in both Fairbanks and Anchorage. Federal subsidies are also an important part of the economy, allowing the state to keep taxes low. Its industrial outputs are crude petroleum, natural gas, coal, gold, precious metals, zinc and other mining, seafood processing, timber and wood products. There is also a growing service and tourism sector. Tourists have contributed to the economy by supporting local lodging.

Largest employers

According to the Alaska Department of Labor and Workforce Development, the following were the state's largest private sector employers in 2010:[47]

BP headquarters in Anchorage.

Alaska Airlines Boeing 737-490 taking off from Ted Stevens Anchorage International Airport.

Fairbanks Memorial Hospital.

Alyeska Prince Hotel in Girdwood.

Rank	Employer name	Average monthly employment in 2010
1	Providence Health & Services	4,000+
2	Walmart/Sam's Club	3,000-3,249
3	Carrs Safeway Alaska Division	2,750-2,999
4	Fred Meyer	2,500-2,749
5	ASRC Energy Services	2,500-2,749
6	Trident Seafoods	2,250-2,499
7	BP Exploration Alaska	2,000-2,249
8	CH2M HILL	1,750-1,999
9	NANA Management Services	1,750-1,999
10	Alaska Native Tribal Health Consortium	1,500-1,749
11	Alaska Airlines	1,500-1,749
12	GCI Communications	1,250-1,499
13	Banner Health (includes Fairbanks Memorial Hospital)	1,250-1,499
14	Southcentral Foundation	1,250-1,499
15	Yukon-Kuskokwim Health Corporation	1,000-1,249
16	FedEx	1,000-1,249
17	ConocoPhillips Alaska	1,000-1,249
18	Alaska USA Federal Credit Union	1,000-1,249
19	United Parcel Service	1,000-1,249
20	McDonald's Restaurants of Alaska	750-999
21	Wells Fargo	750-999
22	Doyon Universal Services	750-999
23	Home Depot	750-999
24	Alaska Regional Hospital	750-999
25	The Alaska Club	750-999
26	Icicle Seafoods	750-999
27	Southeast Alaska Regional Health Consortium	750-999
28	Hope Community Resources	750-999
29	UniSea	750-999
30	Alaska Commercial Company	750-999
31	Costco	750-999
32	Spenard Builders Supply	750-999
33	Lowe's	750-999
34	Alyeska Pipeline Service Company	750-999
35	Alaska Communication Systems	500-749
36	First National Bank Alaska	500-749
37	Central Peninsula Hospital	500-749

38	First Student	500-749
39	Westward Seafood	500-749
40	Mat-Su Regional Medical Center	500-749
41	Alaska Consumer Direct Personal Care	500-749
42	Tanana Chiefs Conference	500-749
43	PeterPan Seafoods	500-749
44	Udelhoven Oilfield System Services	500-749
45	Job Ready/ReadyCare	500-749
46	Schlumberger Technologies	500-749
47	Maniilaq Association	500-749
48	Alaska Hotel Properties/Princess Hotels	500-749
49	Alyeska Resort (includes O'Malley's on the Green)	500-749
50	Ocean Beauty Seafoods	250-499

Energy

Alaska has vast energy resources. Major oil and gas reserves are found in the Alaska North Slope (ANS) and Cook Inlet basins. According to the Energy Information Administration, Alaska ranks second in the nation in crude oil production. Prudhoe Bay on Alaska's North Slope is the highest yielding oil field in the United States and on North America, typically producing about 400000 barrels per day (64000 m^3/d).

The Trans-Alaska Pipeline can transport and pump up to 2.1 million barrels (m^3) of crude oil per day, more than any other crude oil pipeline in the United States. Additionally, substantial coal deposits are found in Alaska's bituminous, sub-bituminous, and lignite coal basins. The United States Geological Survey estimates that there are 85.4 trillion cubic feet (2420 km^3) of undiscovered, technically recoverable gas from natural gas hydrates on the Alaskan North Slope.[48] Alaska also offers some of the highest hydroelectric power potential in the country from its numerous rivers. Large swaths of the Alaskan coastline offer wind and geothermal energy potential as well.[49]

The Trans-Alaska Pipeline transports oil, Alaska's most financially important export, from the North Slope to Valdez. Pertinent are the heat pipes in the column mounts, which disperses heat upwards and prevents melting of permafrost.

Alaska's economy depends heavily on increasingly expensive diesel fuel for heating, transportation, electric power and light. Though wind and hydroelectric power are abundant and underdeveloped, proposals for state-wide energy systems (e.g. with special low-cost electric interties) were judged uneconomical (at the time of the report, 2001) due to low (<$0.50/Gal) fuel prices, long distances and low population.[50] The cost of a US gallon of gas in urban Alaska today is usually $0.30–$0.60 higher than the national average; prices in rural areas are generally significantly higher but vary widely depending on transportation costs, seasonal usage peaks, nearby petroleum development infrastructure and many other factors.

Alaska accounts for one-fifth (20 percent) of domestically produced United States oil production. Prudhoe Bay (North America's largest oil field) alone accounts for 8% of the U.S. domestic oil production.

Permanent Fund

The Alaska Permanent Fund is a constitutionally authorized appropriation of oil revenues, established by voters in 1976 to manage a surplus in state petroleum revenues from oil, largely in anticipation of same from the recently constructed Trans-Alaska Pipeline System. The fund was originally proposed by Governor Keith Miller on the eve of the 1969 Prudhoe Bay lease sale, out of fear that the legislature would spend the entire proceeds of the sale (which amounted to $900 million (US)) at once, and was later championed by Governor Jay Hammond and Kenai state representative Hugh Malone. It has served as an attractive political prospect ever since, diverting revenues which would normally be deposited into the general fund.

The Alaska Constitution was written so as to discourage dedicating state funds for a particular purpose. The Permanent Fund has become the rare exception to this, mostly due to the political climate of distrust existing during the time of its creation. From its initial principal of $734,000, the fund has grown to $40 billion as a result of oil royalties and capital investment programs.[51] Most if not all the principal is invested conservatively outside Alaska. This has led to frequent calls by Alaskan politicians for the Fund to make investments within Alaska, though such a stance has never really gained momentum.

Starting in 1982, dividends from the fund's annual growth have been paid out each year to eligible Alaskans, ranging from an initial $1,000.00 in 1982 (equal to three years' payout, as the distribution of payments was held up in a lawsuit over the distribution scheme) to $3,269.00 in 2008 (which included a one-time $1,200.00 "Resource Rebate"). Every year, the state legislature takes out 8 percent from the earnings, puts 3 percent back into the principal for inflation proofing, and the remaining 5 percent is distributed to all qualifying Alaskans. To qualify for the Permanent Fund Dividend, one must have lived in the state for a minimum of 12 months, maintain constant residency subject to allowable absences,[52] and not be subject to court judgments or criminal convictions which fall under various disqualifying classifications or may subject the payment amount to civil garnishment.

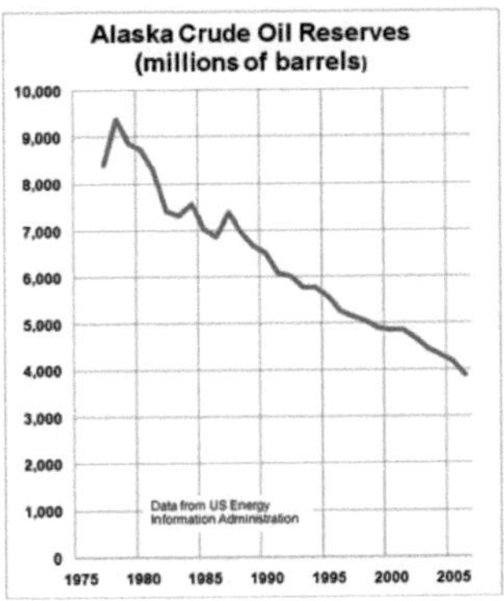

Alaska oil reserves peaked in 1978 and have declined 60% thereafter.

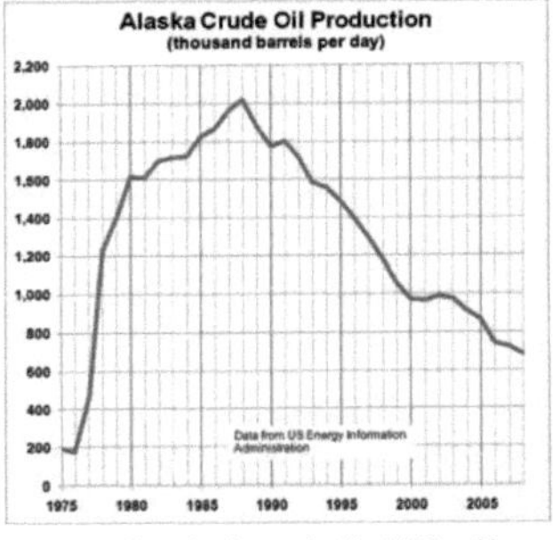

Alaska oil production peaked in 1988 and has declined 65% since.

Cost of living

The cost of goods in Alaska has long been higher than in the contiguous 48 states. This has changed for the most part in Anchorage and to a lesser extent in Fairbanks, where the cost of living has dropped somewhat in the past five years. Federal government employees, particularly United States Postal Service (USPS) workers and active-duty military members, receive a Cost of Living Allowance usually set at 25% of base pay because, while the cost of living has gone down, it is still one of the highest in the country.

The introduction of big-box stores in Anchorage, Fairbanks, and Juneau also did much to lower prices. Wal-Mart opened their first Anchorage store in 1993, and debuted in Fairbanks in 2004. The company currently has locations covering most of the population centers of Alaska, including Juneau, Ketchikan and Kodiak. However, rural Alaska suffers from extremely high prices for food and consumer goods, compared to the rest of the country due to the relatively limited transportation infrastructure. Many rural residents come into these cities and purchase food and goods in bulk from warehouse clubs like Costco and Sam's Club. Some have embraced the free shipping offers[53] of some online retailers to purchase items much more cheaply than they could in their own communities, if they are available at all.

Agriculture

Due to the northern climate and steep terrain, relatively little farming occurs in Alaska. Most farms are in either the Matanuska Valley, about 40 miles (64 km) northeast of Anchorage, or on the Kenai Peninsula, about 60 miles (97 km) southwest of Anchorage. The short 100-day growing season limits the crops that can be grown, but the long sunny summer days make for productive growing seasons. The primary crops are potatoes, carrots, lettuce, and cabbage. The Delta Junction area, about 100 miles (160 km) southeast of Fairbanks, also has a sizable concentration of farms, which mostly lie north and east of Fort Greely. This area was largely set aside and developed under a state program spearheaded by Hammond during his second term as governor. Delta-area crops consist predominately of barley and hay.

Alaska, with no counties, also lacks county fairs. However, a small assortment of state and local fairs (with the Alaska State Fair in Palmer the largest), are held mostly in the late summer. The fairs are mostly located in communities with historic or current agricultural activity, and feature local farmers exhibiting produce in addition to more high-profile commercial activities such as carnival rides, concerts and food. "Alaska Grown" is used as an agricultural slogan.

Halibut is important to the state's economy as both a commercial and sport-caught fish.

Alaska has an abundance of seafood, with the primary fisheries in the Bering Sea and the North Pacific, and seafood is one of the few food items that is often cheaper within the state than outside it. Many Alaskans take advantage of salmon seasons to harvest portions of their household diet while fishing for subsistence, as well as sport. This includes fish taken by hook, net or wheel.[54]

Hunting for subsistence, primarily caribou, moose, and Dall sheep is still common in the state, particularly in remote Bush communities. An example of a traditional native food is Akutaq, the Eskimo ice cream, which can consist of reindeer fat, seal oil, dried fish meat and local berries.

Alaska's reindeer herding is concentrated on Seward Peninsula where wild caribou can be prevented from mingling and migrating with the domesticated reindeer.[55]

Most food in Alaska is transported into the state from "Outside", and shipping costs make food in the cities relatively expensive. In rural areas, subsistence hunting and gathering is an essential activity because imported food is prohibitively expensive. The cost of importing food to villages begins at 7¢ per pound (15¢/kg) and rises rapidly to 50¢ per pound ($1.10/kg) or more. The cost of delivering a 1 US gallon (3.8 L) of milk is about $3.50 in many villages where per capita income can be $20,000 or less. Fuel cost can exceed $8.00 per gallon.

Transportation

Roads

Alaska has few road connections compared to the rest of the U.S. The state's road system covers a relatively small area of the state, linking the central population centers and the Alaska Highway, the principal route out of the state through Canada. The state capital, Juneau, is not accessible by road, only a car ferry, which has spurred several debates over the decades about moving the capital to a city on the road system, or building a road connection from Haines. The western part of Alaska has no road system connecting the communities with the rest of Alaska.

Sterling Highway.

One unique feature of the Alaska Highway system is the Anton Anderson Memorial Tunnel, an active Alaska Railroad tunnel recently upgraded to provide a paved roadway link with the isolated community of Whittier on Prince William Sound to the Seward Highway about 50 miles (80 km) southeast of Anchorage at Portage. At 2.5 miles (4.0 km) the tunnel was the longest road tunnel in North America until 2007.[56] The tunnel is the longest combination road and rail tunnel in North America.

The Susitna River bridge on the Denali Highway is 1036 feet (316 m) long.

Rail

Built around 1915, the Alaska Railroad (ARR) played a key role in the development of Alaska through the 20th century. It links north Pacific shipping through providing critical infrastructure with tracks that run from Seward to Interior Alaska by way of South Central Alaska, passing through Anchorage, Eklutna, Wasilla, Talkeetna, Denali, and Fairbanks, with spurs to Whittier, Palmer and North Pole. The cities, towns, villages, and region served by ARR tracks are known statewide as "The Railbelt". In recent years, the ever-improving paved highway system began to eclipse the railroad's importance in Alaska's economy.

Alaska Railroad "Glacier Discovery" train.

The railroad, though famed for its summertime tour passenger service, played a vital role in Alaska's development, moving freight into Alaska while transporting natural resources southward (i.e., coal from the Usibelli coal mine near Healy to Seward and gravel from the Matanuska Valley to Anchorage).

The Alaska Railroad was one of the last railroads in North America to use cabooses in regular service and still uses them on some gravel trains. It continues to offer one of the last flag stop routes in the country. A stretch of about 60 miles (100 km) of track along an area north of Talkeetna remains inaccessible by road; the railroad provides the only transportation to rural homes and cabins in the area; until construction of the Parks Highway in the 1970s, the railroad provided the only land access to most of the region along its entire route.

In northern Southeast Alaska, the White Pass and Yukon Route also partly runs through the state from Skagway northwards into Canada (British Columbia and Yukon Territory), crossing the border at White Pass Summit. This line is now mainly used by tourists, often arriving by cruise liner at Skagway. It featured in the 1983 BBC television series Great Little Railways.

The Alaska Rail network is not connected to Outside. In 2000, the U.S. Congress authorized $6 million to study the feasibility of a rail link between Alaska, Canada, and the lower 48.[57] [58] [59]

Alaska Rail Marine provides car float service between Whittier and Seattle.

Marine transport

Many cities, towns and villages in the state do not have road or highway access; the only modes of access involve travel by air, river, or the sea.

Alaska's well-developed state-owned ferry system (known as the Alaska Marine Highway) serves the cities of southeast, the Gulf Coast and the Alaska Peninsula. The ferries transport vehicles as well as passengers. The system also operates a ferry service from Bellingham, Washington and Prince Rupert, British Columbia in Canada through the Inside Passage to Skagway. The Inter-Island Ferry Authority also serves as an important marine link for many communities in the Prince of Wales Island region of Southeast and works in concert with the Alaska Marine Highway.

The M/V *Tustumena* (named after Tustumena Glacier) is one of the state's many ferries, providing service between the Kenai Peninsula and Kodiak Island.

In recent years, cruise lines have created a summertime tourism market, mainly connecting the Pacific Northwest to Southeast Alaska and, to a lesser degree, towns along Alaska's gulf coast. The population of Ketchikan may rise by over 10,000 people on many days during the summer, as up to four large cruise ships at a time can dock, debarking thousands passengers.

Air transport

Cities not served by road, sea, or river can be reached only by air, foot, dogsled, or snowmachine accounting for Alaska's extremely well developed bush air services—an Alaskan novelty. Anchorage itself, and to a lesser extent Fairbanks, are served by many major airlines. Because of limited highway access, air travel remains the most efficient form of transportation in and out of the state. Anchorage recently completed extensive remodeling and construction at Ted Stevens Anchorage International Airport to help accommodate the upsurge in tourism (in 2000–2001, the latest year for which data is available, 2.4 million total arrivals to Alaska were counted, 1.7 million by air travel; 1.4 million were visitors).[60] [61]

Alaska Airlines Boeing 737 airliner

Regular flights to most villages and towns within the state that are commercially viable are challenging to provide, so they are heavily subsidized by the federal government through the Essential Air Service program. Alaska Airlines is the only major airline offering in-state travel with jet service (sometimes in combination cargo and passenger Boeing 737-400s) from Anchorage and Fairbanks to regional hubs like Bethel, Nome, Kotzebue, Dillingham, Kodiak, and other larger communities as well as to major Southeast and Alaska Peninsula communities.

The bulk of remaining commercial flight offerings come from small regional commuter airlines such as Era Aviation, PenAir, and Frontier Flying Service. The smallest towns and villages must rely on scheduled or chartered bush flying services using general aviation aircraft such as the Cessna Caravan, the most popular aircraft in use in the state. Much of this service can be attributed to the Alaska bypass mail program which subsidizes bulk mail delivery to Alaskan rural communities. The program requires 70% of that subsidy to go to carriers who offer passenger service to the communities.

Many communities have small air taxi services. These operations originated from the demand for customized transport to remote areas. Perhaps the most quintessentially Alaskan plane is the bush seaplane. The world's busiest seaplane base is Lake Hood, located next to Ted Stevens Anchorage International Airport, where flights bound for remote villages without an airstrip carry passengers, cargo, and many items from stores and warehouse clubs. Alaska has the highest number of pilots per capita of any U.S. state: out of the estimated 663,661 residents, 8,550 are pilots, or about one in 78.[62]

Other transport

Another Alaskan transportation method is the dogsled. In modern times (that is, any time after the mid-late 1920s), dog mushing is more of a sport than a true means of transportation. Various races are held around the state, but the best known is the Iditarod Trail Sled Dog Race, a 1150-mile (1850 km) trail from Anchorage to Nome (although the distance varies from year to year, the official distance is set at 1049 miles (1688 km)). The race commemorates the famous 1925 serum run to Nome in which mushers and dogs like Togo and Balto took much-needed medicine to the diphtheria-stricken community of Nome when all other means of transportation had failed. Mushers from all over the world come to Anchorage each March to compete for cash, prizes, and prestige. The "Serum Run" is another sled dog race that more accurately follows the route of the famous 1925 relay, leaving from the community of Nenana (southwest of Fairbanks) to Nome.[63]

In areas not served by road or rail, primary transportation in summer is by all-terrain vehicle and in winter by snowmobile or "snow machine," as it is commonly referred to in Alaska.

Law and government

State government

Like all other U.S. states, Alaska is governed as a republic, with three branches of government: an executive branch consisting of the Governor of Alaska and the other independently elected constitutional officers; a legislative branch consisting of the Alaska House of Representatives and Alaska Senate; and a judicial branch consisting of the Alaska Supreme Court and lower courts.

The state of Alaska employs approximately 15,000 employees statewide.[64]

The Alaska Legislature consists of a 40-member House of Representatives and a 20-member Senate. Senators serve four year terms and House members two. The Governor of Alaska serves four-year terms. The lieutenant governor runs separately from the governor in the primaries, but during the general election, the nominee for governor and nominee for lieutenant governor run together on the same ticket.

The center of state government in Juneau. The large buildings in the background are, from left to right: the Court Plaza Building (colloquially known as the "Spam Can"), the State Office Building (behind), the Alaska Office Building, the John H. Dimond State Courthouse, and the Alaska State Capitol. Many of the smaller buildings in the foreground are also occupied by state government agencies.

Alaska's court system has four levels: the Alaska Supreme Court, the court of appeals, the superior courts and the district courts.[65] The superior and district courts are trial courts. Superior courts are courts of general jurisdiction, while district courts only hear certain types of cases, including misdemeanor criminal cases and civil cases valued up to $100,000.[65] The Supreme Court and the Court Of Appeals are appellate courts. The Court Of Appeals is required to hear appeals from certain lower-court decisions, including those regarding criminal prosecutions, juvenile delinquency, and habeas corpus.[65] The Supreme Court hears civil appeals and may in its discretion hear criminal appeals.[65]

State politics

Further information: Political party strength in Alaska, Alaska political corruption probe

Although Alaska entered the union as a Democratic state, since the early 1970s Alaska has been characterized as a Republican-leaning state.[66] Local political communities have often worked on issues related to land use development, fishing, tourism, and individual rights. Alaska Natives, while organized in and around their communities, have been active within the Native corporations. These have been given ownership over large tracts of land, which require stewardship.

Alaska is the only state in which possession of one ounce or less of marijuana in one's home is completely legal under state law, though the federal law remains in force.[67]

The state has an independence movement favoring a vote on secession from the United States, with the Alaskan Independence Party labeled as one of "the most significant state-level third parties operating in the 20th century".[68]

Six Republicans and four Democrats have served as governor of Alaska. In addition, Republican Governor Wally Hickel was elected to the office for a second term in 1990 after leaving the Republican party and briefly joining the Alaskan Independence Party ticket just long enough to be reelected. He subsequently officially rejoined the Republican party in 1994.

Taxes

To finance state government operations, Alaska depends primarily on petroleum revenues and federal subsidies. This allows it to have the lowest individual tax burden in the United States,[69] and be one of only five states with no state sales tax, one of seven states that do not levy an individual income tax, and one of two states that has neither. The Department of Revenue Tax Division[70] reports regularly on the state's revenue sources. The Department also issues an annual summary of its operations, including new state laws that directly affect the tax division.

While Alaska has no state sales tax, 89 municipalities collect a local sales tax, from 1–7.5%, typically 3–5%. Other local taxes levied include raw fish taxes, hotel, motel, and bed-and-breakfast 'bed' taxes, severance taxes, liquor and tobacco taxes, gaming (pull tabs) taxes, tire taxes and fuel transfer taxes. A part of the revenue collected from certain state taxes and license fees (such as petroleum, aviation motor fuel, telephone cooperative) is shared with municipalities in Alaska.

Fairbanks has one of the highest property taxes in the state as no sales or income taxes are assessed in the Fairbanks North Star Borough (FNSB). A sales tax for the FNSB has been voted on many times, but has yet to be approved, leading law makers to increase taxes dramatically on other goods such as liquor and tobacco.

In 2008 the Tax Foundation ranked Alaska as having the 4th most "business friendly" tax policy. More "friendly" states were Wyoming, Nevada, and South Dakota.[71]

Federal politics

Presidential elections results

Year	Republican	Democratic
2008	**59.49%** *192,631*	37.83% *122,485*
2004	**61.07%** *190,889*	35.52% *111,025*
2000	**58.62%** *167,398*	27.67% *79,004*
1996	**50.80%** *122,746*	33.27% *80,380*
1992	**39.46%** *102,000*	30.29% *78,294*
1988	**59.59%** *119,251*	36.27% *72,584*
1984	**66.65%** *138,377*	29.87% *62,007*
1980	**54.35%** *86,112*	26.41% *41,842*
1976	**57.90%** *71,555*	35.65% *44,058*
1972	**58.13%** *55,349*	34.62% *32,967*
1968	**45.28%** *37,600*	42.65% *35,411*
1964	34.09% *22,930*	**65.91%** *44,329*
1960	**50.94%** *30,953*	49.06% *29,809*

In presidential elections, the state's electoral college votes have been won by the Republican nominee in every election since statehood, except for 1964. No state has voted for a Democratic presidential candidate fewer times. Alaska supported Democratic nominee Lyndon B. Johnson in the landslide year of 1964, and the 1960 and 1968 elections were close. Since 1972, however, Republicans have carried the state by large margins. In 2008, Republican John McCain defeated Democrat Barack Obama in Alaska, 59.49% to 37.83%. McCain's running mate was Sarah Palin, the state's governor and the first Alaskan on a major party ticket.

The Alaska Bush, central Juneau, midtown and downtown Anchorage, and the area surrounding the University of Alaska Fairbanks campus have been strongholds of the Democratic Party. Matanuska-Susitna Borough and South

Anchorage typically have the strongest Republican showing. As of 2004, well over half of all registered voters have chosen "Non-Partisan" or "Undeclared" as their affiliation,[72] despite recent attempts to close primaries.

Because of its population relative to other U.S. states, Alaska has only one member in the U.S. House of Representatives. This seat is currently being held by Republican Don Young, who was re-elected to his 19th consecutive term in 2008. Alaska's At-large congressional district is currently the world's second-largest legislative constituency by area, behind only the Canadian territory of Nunavut.

In 2008, Governor Sarah Palin became the first Republican woman to run on a national ticket when she became John McCain's Vice Presidential running mate. She continued to be a prominent national figure even after resigning from the governor's job in July 2009.

Alaska's United States Senators belong to Class 2 and Class 3. In 2008, Democrat Mark Begich, mayor of Anchorage, defeated long-time Republican senator Ted Stevens. Stevens had been convicted on seven felony counts of failing to report gifts on Senate financial discloser forms one week before the election. The conviction was set aside in April 2009 after evidence of prosecutorial misconduct emerged.

Republican Frank Murkowski held the state's other senatorial position. After being elected governor in 2002, he resigned from the Senate and appointed his daughter, State Representative Lisa Murkowski as his successor. She won a full six-year term in 2004 and 2010.

Cities, towns and boroughs

Alaska is not divided into counties, as most of the other U.S. states, but it is divided into *boroughs*. Many of the more densely populated parts of the state are part of Alaska's 16 boroughs, which function somewhat similarly to counties in other states. However, unlike county-equivalents in the other 49 states, the boroughs do not cover the entire land area of the state. The area not part of any borough is referred to as the Unorganized Borough.

Anchorage, Alaska's largest city.

The Unorganized Borough has no government of its own, but the U.S. Census Bureau in cooperation with the state divided the Unorganized Borough into 11 census areas solely for the purposes of statistical analysis and presentation. A *recording district* is a mechanism for administration of the public record in Alaska. The state is divided into 34 recording districts which are centrally administered under a State Recorder. All recording districts use the same acceptance criteria, fee schedule, etc., for accepting documents into the public record.

Fairbanks.

Whereas many U.S. states use a three-tiered system of decentralization—state/county/township—most of Alaska uses only two tiers—state/borough. Owing to the low population density, most of the land is located in the Unorganized Borough which, as the name implies, has no intermediate borough government of its own, but is administered directly by the state government. Currently (2000 census) 57.71% of Alaska's area has this status, with 13.05% of the population.

For statistical purposes the United States Census Bureau divides this territory into census areas. Anchorage merged the city government with the Greater Anchorage Area Borough in 1975 to form the

Municipality of Anchorage, containing the city proper and the communities of Eagle River, Chugiak, Peters Creek, Girdwood, Bird, and Indian. Fairbanks has a separate borough (the Fairbanks North Star Borough) and municipality (the City of Fairbanks).

The state's most populous city is Anchorage, home to 278,700 people in 2006, 225,744 of whom live in the urbanized area. The richest location in Alaska by per capita income is Halibut Cove ($89,895). Yakutat City, Sitka, Juneau, and Anchorage are the four largest cities in the U.S. by area.

Alaska's capital city, Juneau.

Cities of 100,000 or more people

- Anchorage

Cities and census-designated places cf 10,000–100,000 people

- Fairbanks
- Juneau (State Capital)
- Knik-Fairview
- College

Cities and census-designated places of 1,000–10,000 people

The Homer Spit in Homer.

Sitka	Homer	Eielson AFB	Lazy Mountain
Ketchikan	Farmers Loop	Ester	Sutton-Alpine
Wasilla	Fishhook	Wrangell	Metlakatla
Kenai	Nikiski	Dillingham	Cohoe
Lakes	Unalaska	Deltana	Kodiak Station
Tanaina	Barrow	Cordova	Susitna North
Kalifornsky	Soldotna	Prudhoe Bay	Tok
Meadow Lakes	Valdez	North Pole	Craig
Steele Creek	Nome	Willow	Diamond Ridge
Kodiak	Goldstream	Ridgeway	Salcha
Bethel	Big Lake	Bear Creek	Hooper Bay
Palmer	Butte	Fritz Creek	Farm Loop
Chena Ridge	Kotzebue	Anchor Point	Akutan
Sterling	Petersburg	Houston	Healy
Gateway	Seward	Haines	

Smaller towns

Alaska has many smaller towns, especially in the Alaska Bush, which are generally inaccessible by road.

Education

The Kachemak Bay campus of the University of Alaska Anchorage.

The Alaska Department of Education and Early Development administers many school districts in Alaska. In addition, the state operates a boarding school, Mt. Edgecumbe High School in Sitka; and provides partial funding for other boarding schools, including Nenana Student Living Center in Nenana and The Galena Interior Learning Academy in Galena.[73]

There are more than a dozen colleges and universities in Alaska. Accredited universities in Alaska include the University of Alaska Anchorage, University of Alaska Fairbanks, University of Alaska Southeast, and Alaska Pacific University.[74]

The Alaska Department of Labor and Workforce Development operates AVTEC, Alaska's Institute of Technology [75]. Campuses in Seward and Anchorage offer 1 week to 11 month training programs in areas as diverse as Information Technology, Welding, Nursing, and Mechanics.

Alaska has had a problem with a "brain drain". Many of its young people, including most of the highest academic achievers, leave the state after high school graduation and do not return. The University of Alaska has attempted to combat this by offering partial four-year scholarships to the top 10% of Alaska high school graduates, via the Alaska Scholars Program.[76]

Public health and public safety

The Alaska State Troopers are Alaska's statewide police force. They have a long and storied history, but were not an official organization until 1941. Before the force was officially organized, law enforcement in Alaska was handled by various federal agencies. Larger towns usually have their own local police and some villages rely on "Public Safety Officers" who have police training but do not carry firearms. In much of the state, the troopers serve as the only police force available. In addition to enforcing traffic and criminal law, wildlife Troopers enforce hunting and fishing regulations. Due to the varied terrain and wide scope of the Troopers' duties, they employ a wide variety of land, air, and water patrol vehicles.

Many rural communities in Alaska are considered "dry," having outlawed the importation of alcoholic beverages .[77] Suicide rates for rural residents are higher than urban.[78]

Domestic abuse and other violent crimes are also at high levels in the state; this is in part linked to alcohol abuse.[79]

Alaska also has the highest rate of sexual assault in the nation. The average age of sexually assaulted victims is 16 years old. In four out of five cases, the suspects were relatives, friends or acquaintances.[80]

Culture

Some of Alaska's popular annual events are the Iditarod Trail Sled Dog Race that starts in Anchorage and ends in Nome, World Ice Art Championships in Fairbanks, the Alaska Hummingbird Festival in Ketchikan, the Sitka Whale Fest, and the Stikine River Garnet Fest in Wrangell. The Stikine River features the largest springtime concentration of American Bald Eagles in the world.

The Alaska Native Heritage Center celebrates the rich heritage of Alaska's 11 cultural groups. Their purpose is to enhance self-esteem among Native people and to encourage cross-cultural exchanges among all people. The Alaska Native Arts Foundation promotes and markets Native art from all regions and cultures in the State, both on

A dog team in the Iditarod race, arguably the most popular winter event in Alaska.

the internet; at its gallery in Anchorage, 500 West Sixth Avenue, and at the Alaska House New York [81], 109 Mercer Street in SoHo.[82]

Music

Influences on music in Alaska include the traditional music of Alaska Natives as well as folk music brought by later immigrants from Russia and Europe. Prominent musicians from Alaska include singer Jewel, traditional Aleut flautist Mary Youngblood, folk singer-songwriter Libby Roderick, Christian music singer/songwriter Lincoln Brewster, metal/post hardcore band 36 Crazyfists and the groups Pamyua and Portugal. The Man.

There are many established music festivals in Alaska, including the Alaska Folk Festival, the Fairbanks Summer Arts Festival, the Anchorage Folk Festival [83], the Athabascan Old-Time Fiddling Festival, the Sitka Jazz Festival, and the Sitka Summer Music Festival. The most prominent orchestra in Alaska is the Anchorage Symphony Orchestra, though the Fairbanks Symphony Orchestra and Juneau Symphony are also notable. The Anchorage Opera is currently the state's only professional opera company, though there are several volunteer and semi-professional organizations in the state as well.

The official state song of Alaska is "Alaska's Flag", which was adopted in 1955; it celebrates the flag of Alaska.

Movies filmed in Alaska

Films featuring Alaskan wolves usually employ domesticated wolf-dog hybrids to stand in for wild wolves.

Alaska's first independent picture all made on place was in the silent years. *The Chechahcos* was produced by Alaskan businessman Austin E. Lathrop and filmed in and around Anchorage. It was released in 1924 by the Alaska Moving Picture Corporation and was the only film the company made.

One of the most prominent movies filmed in Alaska is MGM's *Eskimo/Mala The Magnificent*, starring Alaska Native Ray Mala. In 1932 an expedition set out from MGM's studios in Hollywood to Alaska to film what was then billed as "The Biggest Picture Ever Made." Upon arriving in Alaska, they set up "Camp Hollywood" in Northwest Alaska, where they lived during the duration of the filming. Louis B. Mayer spared no expense in spite of the remote location, going so far as to hire the chef from the Hotel Roosevelt in Hollywood to prepare meals. When *Eskimo* premiered at the Astor Theatre in New York City, the studio received the largest amount of feedback in its history to that point. *Eskimo* was critically acclaimed and released worldwide; as a result, Mala became an international movie star. *Eskimo* won the first Oscar for Best Film Editing at the Academy Awards, and was also responsible for showcasing and preserving aspects of Inupiat culture on film.

The 1983 Disney movie *Never Cry Wolf* was at least partially shot in Alaska. The 1991 film *White Fang*, starring Ethan Hawke, was filmed in and around Haines, Alaska. Steven Seagal's 1994 *On Deadly Ground*, starring Michael Caine, was filmed in part at the Worthington Glacier near Valdez.[84] The 1999 John Sayles film *Limbo*, starring David Strathairn, Mary Elizabeth Mastrantonio, and Kris Kristofferson, was filmed in Juneau.

The psychological thriller *Insomnia*, starring Al Pacino and Robin Williams, was shot in Canada, but was set in Alaska. The 2007 horror feature *30 Days of Night* is set in Barrow, Alaska, but was filmed in New Zealand. Most films and television shows set in Alaska are not filmed there; for example, *Northern Exposure*, set in the fictional town of Cicely, Alaska, was actually filmed in Roslyn, Washington.

The 2007 film directed by Sean Penn, *Into The Wild*, was partially filmed and set in Alaska. The film, which is based on the novel of the same name, follows the adventures of Christopher McCandless, who died in a remote abandoned bus along the Stampede Trail west of Healy in 1992.

State symbols

- State Motto: North to the Future
- Nicknames: "The Last Frontier" or "Land of the Midnight Sun" or "Seward's Icebox"
- State bird: Willow Ptarmigan, adopted by the Territorial Legislature in 1955. It is a small (15–17 inches) Arctic grouse that lives among willows and on open tundra and muskeg. Plumage is brown in summer, changing to white in winter. The Willow Ptarmigan is common in much of Alaska.
- State fish: King Salmon, adopted 1962.
- State flower: wild/native Forget-Me-Not, adopted by the Territorial Legislature in 1917.[85] It is a perennial that is found throughout Alaska, from Hyder to the Arctic Coast, and west to the Aleutians.

- State fossil: Woolly Mammoth, adopted 1986.
- State gem: Jade, adopted 1968.
- State insect: Four-spot skimmer dragonfly, adopted 1995.
- State land mammal: Moose, adopted 1998.
- State marine mammal: Bowhead Whale, adopted 1983.
- State mineral: Gold, adopted 1968.
- State song: "Alaska's Flag"
- State sport: Dog Mushing, adopted 1972.
- State tree: Sitka Spruce, adopted 1962.
- State dog: Alaskan Malamute, adopted 2010.[86]
- State soil: Tanana,[87] adopted unknown.

The Forget-me-not is the state's official flower
and bears the same blue and gold as the state flag

See also

- List of National Register of Historic Places in Alaska
- List of people from Alaska
- List of places in Alaska
- U.S. state

References

[1] "Language Legislation in the U.S.A." (http://web.archive.org/web/20090426134135/http://ourworld.compuserve.com/homepages/ JWCRAWFORD/langleg.htm). CompuServe.com. Archived from the original (http://ourworld.compuserve.com/homepages/jwcrawford/ langleg.htm) on 2009-04-26. .

[2] "RCN – Digital Cable TV, High-Speed Internet Service & Phone in Boston, Chicago, New York City, Philadelphia, Washington, D.C. and the Lehigh Valley" (http://users.rcn.com/crawj/kritz.htm). Users.rcn.com. . Retrieved 2010-06-02.

[3] "Resident Population Data – 2010 Census" (http://2010.census.gov/2010census/data/apportionment-pop-text.php). United States Census Bureau. . Retrieved 2011-02-27.

[4] "Elevations and Distances in the United States" (http://egsc.usgs.gov/isb/pubs/booklets/elvadist/elvadist.html). United States Geological Survey. 2001. . Retrieved October 21, 2011.

[5] "Census Bureau" (http://factfinder.census.gov/servlet/DTTable?_bm=y&-geo_id=04000US02&-ds_name=PEP_2007_EST& -mt_name=PEP_2007_EST_G2007_T001). Factfinder.census.gov. . Retrieved 2010-06-02.

[6] Ransom, J. Ellis. 1940. *Derivation of the Word "Alaska"*. American Anthropologist n.s., 42: pp. 550–551

[7] Benson, Carl (1998-09-02). "Alaska's Size in Perspective" (http://www.gi.alaska.edu/ScienceForum/ASF14/1404.html). Geophysical Institute, University of Alaska Fairbanks. . Retrieved 2007-11-19.

[8] The other three exclaves of the United States are the Northwest Angle of Minnesota, Point Roberts, Washington and Alburgh, Vermont.

[9] Alaska.com. "Alaska.com" (http://www.alaska.com/regions/). Alaska.com. . Retrieved 2010-06-02.

[10] Porco, Peter (June 23, 2003). "Long said to be second to Fundy, city tides aren't even close". *Anchorage Daily News*: A1.

[11] "Alaska Hydrology Survey" (http://www.dnr.state.ak.us/mlw/water/hydro/). Division of Mining, Land, and Water; Alaska Department of Natural Resources. .

[12] "Alaska Facts" (http://web.archive.org/web/20080228134807/http://www.knls.org/English/akfact.htm). Knls.org. Archived from the original (http://www.knls.org/English/akfact.htm) on February 28, 2008. . Retrieved 2010-06-02.

[13] "Western States Data Public Land Acreage" (http://www.wildlandfire.com/docs/2007/western-states-data-public-land.htm). Wildlandfire.com. 2007-11-13. . Retrieved 2010-06-02.

[14] Mean Annual Precipitation in Alaska-Yukon (http://www.ocs.orst.edu/pub/maps/Precipitation/Total/States/AK/ak_ppt.gif). Oregon Climate Service at Oregon State University. Retrieved October 23, 2006.

[15] "NOAA Weather Radio All Hazards Information — Alaska Weather Interesting Facts and Records" (http://www.arh.noaa.gov/docs/ AKWXfacts.pdf) (PDF). National Oceanic and Atmospheric Administration. . Retrieved 2007-01-03.

[16] "State Extremes" (http://www.wrcc.dri.edu/htmlfiles/state.extremes.html). Western Regional Climate Center, Desert Research Institute. . Retrieved 2007-01-03.

[17] NOAA – National Oceanic and Atmospheric Administration "SD Weather History and Trivia for May: May 1" (http://www.crh.noaa. gov/fsd/?n=fsdtrivia05). National Oceanic and Atmospheric Administration. NOAA – National Oceanic and Atmospheric Administration.

Retrieved 2007-01-03.

[18] "FAQ ALASKA — Frequently Asked Questions About Alaska: Weather" (http://sled.alaska.edu/akfaq/aksuper.html#wea). Statewide Library Electronic Doorway, University of Alaska Fairbanks. 2005-01-17. . Retrieved 2007-01-03.

[19] Ned Rozell (2003-01-23). "The Coldest Place in North America" (http://www.gi.alaska.edu/ScienceForum/ASF16/1630.html). Geophysical Institute of the University of Alaska Fairbanks. . Retrieved 2007-01-03.

[20] History for Barrow, Alaska. Monthly Summary for July 2006 (http://www.wunderground.com/history/airport/PABR/2006/7/23/MonthlyHistory.html). Weather Underground. Retrieved October 23, 2006.

[21] Свердлов Л. М. Русское поселение на Аляске в XVII в.? «Природа». М., 1992. № 4. С.67–69.

[22] (http://www.fartuna.ru/go.php?ad=489047)

[23] Аронов В. Н. Патриарх Камчатского мореходства. // «Вопросы истории рыбной промышленности Камчатки»: Историко-краеведческий сб. – Вып. 3. – 2000. Вахрин С. Покорители великого океана. Петроп.-Камч.: Камштат, 1993.

[24] these three Aleutian outer islands are about 460 miles (740 km) away from continental USSR, 920 miles (1480 km) from continental Alaska (U.S.), 950 miles (1530 km) from Japan.

[25] "Resident Population Data – 2010 Census" (http://2010.census.gov/2010census/data/apportionment-pop-text.php). 2010.census.gov. . Retrieved 2011-05-29.

[26] U.S. Census Bureau (2008-12-15). "Cumulative Estimates of the Components of Population Change for the United States, Regions and States: April 1, 2000 to July 1, 2008 (NST-EST2008-04)" (http://www.census.gov/popest/states/tables/NST-EST2008-04.csv) (CSV). . Retrieved 2009-01-16.

[27] "Census Bureau" (http://www.census.gov/population/cen2000/phc-t2/tab01.txt). census.gov. . Retrieved 2010-11-07.

[28] Bls.gov (http://www.bls.gov/lau/); Local Area Unemployment Statistics

[29] "American FactFinder" (http://factfinder2.census.gov/faces/tableservices/jsf/pages/productview.xhtml?pid=DEC_10_PL_QTPL&prodType=table). Factfinder2.census.gov. 2010-10-05. . Retrieved 2011-05-29.

[30] American FactFinder, United States Census Bureau. "Census Bureau" (http://factfinder.census.gov/servlet/ADPTable?_bm=y&-geo_id=04000US02&-qr_name=ACS_2007_3YR_G00_DP3YR2&-ds_name=ACS_2007_3YR_G00_&-_lang=en&-redoLog=false&-_sse=on). Factfinder.census.gov. . Retrieved 2010-06-02.

[31] "Adherents.com" (http://www.adherents.com/Na/Na_472.html). Adherents.com. . Retrieved 2010-06-02.

[32] "Believe it or not, Alaska's one of nation's least religious states" (http://web.archive.org/web/20090116035021/http://www.adn.com/life/story/463303.html). Anchorage Daily News. 2008-07-13. Archived from the original (http://www.adn.com/life/story/463303.html) on 2009-01-16. .

[33] "Religious Affiliations 2000" (http://www.thearda.com/mapsReports/reports/state/02_2000.asp). *Alaska State Membership Report*. Association of Religion Data Archives. . Retrieved 2008-03-31.

[34] "LDS Newsroom Statistical Information" (http://web.archive.org/web/20080730060850/http://newsroom.lds.org/ldsnewsroom/eng/statistical-information). Newsroom.lds.org. Archived from the original (http://newsroom.lds.org/ldsnewsroom/eng/statistical-information) on July 30, 2008. . Retrieved 2010-06-02.

[35] Dixon, Martha (2004-09-12). "*Religious legacy lives on in Alaska*" (http://news.bbc.co.uk/2/hi/americas/3531458.stm). BBC News. . Retrieved 2010-06-02.

[36] "Welcome to SLED: FAQ Alaska" (http://sled.alaska.edu/akfaq/akchron.html). Sled.alaska.edu. . Retrieved 2010-06-02.

[37] "An early Russian Orthodox Church" (http://vilda.alaska.edu/u?/cdmg11,4904). Vilda.alaska.edu. . Retrieved 2010-06-02.

[38] "Association of Religion Data Archive" (http://www.thearda.com/mapsReports/maps/map.asp?state=101&variable=201). Thearda.com. . Retrieved 2010-06-02.

[39] http://www.census.gov/compendia/statab/2011/tables/11s0077.pdf

[40] "First Muslim cemetery opens in Alaska" (http://web.archive.org/web/20090116035850/http://dwb.adn.com/news/alaska/ap_alaska/story/8656236p-8548061c.html). Archived from the original (http://dwb.adn.com/news/alaska/ap_alaska/story/8656236p-8548061c.html) on 2009-01-16. .

[41] "Engaging Muslim: Religion, Culture, Politics" (http://web.archive.org/web/20090215202834/http://engagingmuslims.alaskapacific.edu/). Archived from the original (http://engagingmuslims.alaskapacific.edu/) on 2009-02-15. .

[42] "Alaskan Muslims Avoid Conflict" (http://www.humanitynews.net/index.php/main/more/alaskan_muslims_avoid_conflict/). Humanitynews.net. 2005-07-07. . Retrieved 2010-06-02.

[43] "Mosque milestone for Alaska Muslims – Americas" (http://english.aljazeera.net/video/americas/2010/12/20101225111741183159.html). Al Jazeera English. 2010-12-25. . Retrieved 2011-05-29.

[44] Kalyan, Mala. "Shri Ganesha Mandir of Alaska" (http://web.archive.org/web/20090201151519/http://caia.arctic.us/?Shri_Ganesha_Mandir_of_Alaska). Cultural Association of India Anchorage. Archived from the original (http://www.caia.arctic.us/?Shri_Ganesha_Mandir_of_Alaska) on 2009-02-11. .

[45] "Hindu Temples in USA – Hindu Mandirs in USA" (http://www.hindumandir.us/west-coast.html#AK). Hindumandir.us. . Retrieved 2010-06-02.

[46] "Holi & Baisakhi celebrated by Alaskan Hindus and Sikhs" (http://web.archive.org/web/20090201151459/http://caia.arctic.us/?Holi_/_Baisakhi_Celebration:). Cultural Association of India Anchorage. Archived from the original (http://www.caia.arctic.us/?Holi_/_Baisakhi_Celebration:) on 2009-02-11. .

[47]   " Alaska's 100 Largest Private Sector Employers in 2010 (July 2011 Alaska Economic Trends)." Alaska Department of Labor and Workforce Development (2007). (http://laborstats.alaska.gov/lgstemp/lgstemp.htm)

[48]   "Gas Hydrates on Alaska's North Slope" (http://www.usgs.gov/corecast/details.asp?ep=74). Usgs.gov. . Retrieved 2010-06-02.

[49]   "EIA State Energy Profiles: Alaska" (http://tonto.eia.doe.gov/state/state_energy_profiles.cfm?sid=AK). Tonto.eia.doe.gov. 2009-08-27. . Retrieved 2010-11-07.

[50]   "Screening Report for Alaska Rural Energy Plan" (http://web.archive.org/web/20080216014031/http://www.dced.state.ak.us/dca/ AEIS/PDF_Files/AIDEA_Energy_Screening.pdf). 2001-04. Archived from the original (http://www.dced.state.ak.us/dca/AEIS/ PDF_Files/AIDEA_Energy_Screening.pdf) on 2008-02-16. .

[51]   Alaska Permanent Fund Corporation (http://www.apfc.org/)

[52]   "State of Alaska Permanent Fund Division" (http://www.pfd.state.ak.us/dividendamounts/index.aspx). Pfd.state.ak.us. . Retrieved 2010-06-02.

[53]   FreeShipping.org (http://www.freeshipping.org/category/alaska-free-shipping-coupons/) for examples of companies offering free shipping to Alaska

[54]   "Alaska Department of Fish and Game" (http://www.adfg.alaska.gov/index.cfm?adfg=fishingPersonalUse.main). Adfg.alaska.gov. . Retrieved 2011-05-29.

[55]   "Reindeer Herding" (http://reindeer.salrm.uaf.edu/about_reindeer/seward_peninsula.php). Reindeer.salrm.uaf.edu. . Retrieved 2010-11-07.

[56]   completion of the 3.5 mile (5.6 km) Interstate 93 tunnel as part of the "Big Dig" project in Boston, Massachusetts.

[57]   Barbara Yaffe (2011-01-02). "Alaska Oil / BC Tar sands via rail" (http://communities.canada.com/vancouversun/blogs/powerplay/ archive/2010/12/13/transporting-oil-across-b-c.aspx). . Retrieved 2011-01-02.

[58]   Allan Dowd (2007-06-27). "Economic study touts Alaska-Canada rail link" (http://www.reuters.com/article/idUSN2138860820070622). Reuters. . Retrieved 2011-01-02.

[59]   AlaskaCanadaRail.org (2005-01-02). "Alaska Canada Rail Link" (http://alaskacanadarail.com/index.html). . Retrieved 2011-01-02.

[60]   State of Alaska Office of Economic Development. *Alaska Visitor Arrivals and Profile-Summer 2001* (http://www.commerce.state.ak.us/ dca/toubus/pub/AVSPSummerArrival111402.pdf). November, 2002. Retrieved September 11, 2006.

[61]   State of Alaska Office of Economic Development. *Alaska Visitor Arrivals and Profile-Fall/Winter 2001* (http://www.commerce.state.ak. us/oed/toubus/pub/AVSPWinterArrival111402.pdf). November, 2002. Retrieved September 11, 2006.

[62]   Federal Aviation Administration. *2005 U.S. Civil Airman Statistics* (http://www.faa.gov/data_research/aviation_data_statistics/ civil_airmen_statistics/2005/)

[63]   "Norman Vaughan Serum Run" (http://www.serumrun.org/). United Nations. 2010-04-15. . Retrieved 2010-06-02.

[64]   "State of Alaska" (http://dop.state.ak.us/employeeOrientation/introduction). Dop.state.ak.us. . Retrieved 2010-06-02.

[65]   "About the Alaska Court System" (http://www.state.ak.us/courts/ctinfo.htm). State.ak.us. . Retrieved 2010-06-02.

[66]   "National Journal Alaska State Profile" (http://election.nationaljournal.com/states/ak.htm). Election.nationaljournal.com. . Retrieved 2010-06-02.

[67]   Volz, Matt (2006-07-11). "Judge rules against Alaska marijuana law" (http://seattletimes.nwsource.com/html/localnews/ 2003118645_webpot10.html). *The Seattle Times* (Frank A. Blethen). . Retrieved 2008-05-22.

[68]   Doughtery, J. (2001, February 25). Alaska party stumps for independence. World Net Daily. Retrieved from A Free Press for a Free People (http://www.worldnetdaily.com/news/article.asp?ARTICLE_ID=21840)

[69]   CNN Money (2005). "How tax friendly is your state?" Retrieved from CNN website (http://money.cnn.com/pf/features/lists/ taxesbystate2005/index.html).

[70]   "Department of Revenue Tax Division" (http://www.tax.state.ak.us/). Tax.state.ak.us. . Retrieved 2010-06-02.

[71]   "Educating Taxpayers Since 1937" (http://www.taxfoundation.org/research/topic/11.html). The Tax Foundation. . Retrieved 2010-06-02.

[72]   "State of Alaska" (http://web.archive.org/web/20080225223222/http://www.gov.state.ak.us/ltgov/elections/regbypty.htm). Gov.state.ak.us. Archived from the original (http://www.gov.state.ak.us/ltgov/elections/regbypty.htm) on 2008-02-25. . Retrieved 2010-06-02.

[73]   "Alaska ICE" (http://www.alaskaice.org/material.php?matID=138). Alaska ICE. . Retrieved 2010-06-02.

[74]   These are the only three universities in the state ranked by US News and World Report. (http://www.usnews.com/usnews/edu/college/ rankings/rankindex_brief.php)

[75]   http://avtec.labor.state.ak.us/

[76]   "UA Scholars Program – Frequently Asked Questions" (http://web.archive.org/web/20080309023826/http://www.alaska.edu/ scholars/faq.xml). Archived from the original (http://www.alaska.edu/scholars/faq.xml#scholars_award) on 2008-03-09. . Retrieved 2009-12-28.

[77]   "AK Dept. of Public Safety Alcoholic Beverage Control Board" (http://www.dps.state.ak.us/abc/restrictions.aspx). Dps.state.ak.us. . Retrieved 2010-06-02.

[78]   "State of Alaska" (http://www.hss.state.ak.us/suicideprevention/statistics_pages_sspc/AKsuiciderate_rural_urban_90-00.htm). Hss.state.ak.us. . Retrieved 2010-06-02.

[79]   "Alaska State Troopers FY 2008 Byrne Grant Application" (http://notes5.state.ak.us/pn/pubnotic.nsf/0/ 8dad412f35cf9ba1892574e30001c413/$FILE/JAG+supplemental+program+narrative.pdf) (PDF). . Retrieved 2010-06-02.

[80] D'oro, Rachel (2008-01-30). "Rural Alaska steeped in sexual violence" (http://www.usatoday.com/news/nation/ 2008-01-29-alaska-sexualassaults_N.htm). *USA Today.* . Retrieved 2010-12-31.

[81] http://www.alaskahouseny.org/index.aspx?ModuleID=1

[82] www.alaskanativearts.org

[83] http://anchoragefolkfestival.org/

[84] "On Deadly Ground" (http://www.filminamerica.com/Movies/OnDeadlyGround/). Filminamerica.com. . Retrieved 2010-11-07.

[85] "Alaska Conservation Foundation – State Symbols" (http://web.archive.org/web/20090225094636/http://akcf.org/_pages/ about_ACF/about_alaska/state_symbols.php). Archived from the original (http://www.akcf.org/_pages/about_ACF/about_alaska/ state_symbols.php) on 2009-02-25. .

[86] "It's official: Malamute now Alaska's state dog – KTUU.com I Alaska's news and information source I" (http://www.ktuu.com/Global/ story.asp?S=12473783). KTUU.com. 2010-05-12. . Retrieved 2010-06-02.

[87] TANANA – ALASKA STATE SOIL (ftp://ftp-fc.sc.egov.usda.gov/NSSC/StateSoil_Profiles/ak_soil.pdf) U.S. Department of Agriculture

External links

- Alaska Community Database System (http://www.dced.state.ak.us/dca/commdb/CF_COMDB.htm)
- Alaska's Digital Archives (http://vilda.alaska.edu/index.php)
- Alaska (http://www.alcnet.org/projects/overview/alaska), project area of the American Land Conservancy (http://www.alcnet.org/)
- Alaska Inter-Tribal Council (http://www.aitc.org/)
- Alaska (http://www.dmoz.org/Regional/North_America/United_States/Alaska/) at the Open Directory Project
- Big, Beautiful Alaska (http://www.life.com/gallery/23236/big-beautiful-alaska#index/0) – slideshow by *Life magazine*

U.S. Government

- Energy & Environmental Data for Alaska (http://tonto.eia.doe.gov/state/state_energy_profiles.cfm?sid=AK)
- USGS real-time, geographic, and other scientific resources of Alaska (http://www.usgs.gov/state/state.asp?State=AK)
- US Census Bureau (http://quickfacts.census.gov/qfd/states/02000.html)
- Alaska State Facts (http://www.ers.usda.gov/statefacts/ak.htm)
- Documents on Alaskan Statehood at the Dwight D. Eisenhower Presidential Library (http://eisenhower.archives.gov/Research/Digital_Documents/Alaska/Alaska.html)
- Guide to collections containing information on Alaskan statehood at the Eisenhower Presidential Library (http://eisenhower.archives.gov/Research/Subject_Guides/PDFs/Agriculture.pdf)

State government

- State of Alaska website (http://www.alaska.gov/)
- Alaska State Databases (http://wikis.ala.org/godort/index.php/Alaska) – Annotated list of searchable databases produced by Alaska state agencies and compiled by the Government Documents Roundtable of the American Library Association.
- Alaska Department of Natural Resources, Recorder's Office (http://www.dnr.state.ak.us/ssd/recoff/terminology.cfm)

gag:Alaska mrj:Аляска

Article Sources and Contributors

Denali Highway *Source*: http://en.wikipedia.org/w/index.php?title=Denali_Highway *Contributors*: Aconcagua, Atanamir, Backspace, Beeblebrox, BigOldMooseHunter, CharlotteWebb, Chris the speller, Cobaltcigs, D6, Dough4872, Gcapp1959, Imzadi1979, JKBrooks85, JustAGal, Librsh, LilHelpa, Lpangelrob, Lunawisp, Mangostar, Mild Bill Hiccup, NE2, Nocturnal Wanderer, RPellessier, Rschen7754, SPUI, SamBelton, Theking17825, Wintergrouse, Zero Gravitas, 11 anonymous edits

Richardson Highway *Source*: http://en.wikipedia.org/w/index.php?title=Richardson_Highway *Contributors*: Aconcagua, Atanamir, Backspace, Balloonguy, Beeblebrox, Delirium, Deville, Gcapp1959, Imzadi1979, Jerrykme, Jonathan.s.kt, Lar, Lpangelrob, Marsal20, Master son, Mattingly23, McGehee, NE2, Nikopoley, Onore Baka Sama, RadioKAOS, RedWolf, Rich Farmbrough, SPUI, Steinberger, Tassedethe, Theking17825, Twinpinesmall, Zero Gravitas, 11 anonymous edits

Denali National Park and Preserve *Source*: http://en.wikipedia.org/w/index.php?title=Denali_National_Park_and_Preserve *Contributors*: 127x0x0x1, 78.26, AKGhetto, Ab5602, Abce2, Acroterion, Ahoerstemeier, Akdamien, AlexiusHoratius, Allstarecho, Andytuba, Annie Duffy, Antandrus, Appraiser, ArcticBartek, Averette, Backspace, Ballista, BanyanTree, Barek, Beeblebrox, Bevo, Bhadani, BillC, BillShurts, Bkonrad, BoomerAB, Brambleclawx, Bubba hotep, Buster2058, C.Fred, Calvin 1998, Camw, CanadianLinuxUser, CanisRufus, Carlosguitar, Catapult, Chris Light, Chris the speller, Ckatz, Cobaltcigs, Cpuwhiz11, Cubic, D6, DSchissler, Darklilac, Darwinek, Davejenk1ns, Deanshan, Deor, Dlaz, Droll, Eddz60, Edgar181, Editore99, Eenu, El C, Eoghanacht, Epolk, Er Komandante, Eubulides, Ewlyahoocorn, Fastestdogever, Favonian, Fbdave, Firsfron, Funandtrvl, Funnyhat, Gbrims, Gilliam, Golden Penguin, HennessyC, Hike395, Hmains, Hopper901, Horst, Hottentot, Ish ishwar, Ixfd64, Jarzalbe, Jengod, Jiang, Jllm06, Jmabel, Joey80, Jojhutton, Jpatokal, Jusdafax, Kablammo, Karl-Henner, Kotjze, KrakatoaKatie, LazyLizaJane, Lightmouse, LittleOldMe, Lovewolves, Luckynat4, MDuchek, MONGO, Mac'ero, Masterjamie, Mcghiever, Mesmes, Mike2523, Nehrams2020, Npgallery, Nyttend, Open state, Oreo Priest, Paukrus, Pavelow235, Pepper, Pharaoh of the Wizards, Postdlf, Pschemp, Qblik, Qxz, Ram-Man, RedWolf, Reywas92, Rhatsa26X, Rmhermen, Sbfw, Shannanan, Stemonitis, Takahashi J, Tavix, The Rambling Man, Theodolite, Thingg, Veledan, Virgy, VitaleBaby, Vsmith, Wetman, Wknight94, Wrh2, 211 anonymous edits

George Parks Highway *Source*: http://en.wikipedia.org/w/index.php?title=George_Parks_Highway *Contributors*: Aconcagua, Allen3, Atanamir, Backspace, Christy747, Deanshan, Deirdre, Dv82matt, FCYTravis, Imzadi1979, InNuce, Jerrykme, Ken Gallager, Lpangelrob, Ltljltlj, NE2, RedWolf, SPUI, Shanoman, Skew-t, Specs112, Theking17825, TwinsMetsFan, Vegaswikian, Zero Gravitas, Δ, 14 anonymous edits

Tangle Lakes *Source*: http://en.wikipedia.org/w/index.php?title=Tangle_Lakes *Contributors*: Aconcagua, Backspace, Beeblebrox, Calliopejen1, Grousehawker, Hmains, Jllm06, Jorgenev, Librsh

Copper River (Alaska) *Source*: http://en.wikipedia.org/w/index.php?title=Copper_River_%28Alaska%29 *Contributors*: Aconcagua, AlaskaRiverGuide, AlaskaTrekker, AlphaPyro, Anna Frodesiak, Backspace, BanyanTree, Bobblewik, CanisRufus, Dé, Dankarl, Danny-w, Dcornwall, Decumanus, Dell Adams, Gadfium, Gaiats, Gholton, Golden Penguin, Gr8fl, Hmains, Jimp, Johnbod, LeheckaG, Librsh, Lightmouse, Mattisse, Maximus Rex, Mircea cs, Nocturnal Wanderer, Originalwana, Plasticup, Portsample, Radon210, Rich Farmbrough, Spireguy, T.J.V., Tolanor, Vedrfolnir, Yksin, Zero Gravitas, 23 anonymous edits

Mount Deborah *Source*: http://en.wikipedia.org/w/index.php?title=Mount_Deborah *Contributors*: Backspace, Droll, Hmains, RadioKAOS, Ratagonia, RedWolf, Spireguy, Xurei, 1 anonymous edits

Wrangell Mountains *Source*: http://en.wikipedia.org/w/index.php?title=Wrangell_Mountains *Contributors*: Abune, Aconcagua, AlaskaTrekker, Backspace, BanyanTree, Bryan Derksen, CanisRufus, Caroig, CommonsDelinker, Decumanus, Golden Penguin, Hike395, Hmains, Horst, Jonathan.s.kt, RedWolf, Seattle Skier, Spireguy, Stemonitis, Stylebox, Volcanoguy, Xurei, 8 anonymous edits

Chugach Mountains *Source*: http://en.wikipedia.org/w/index.php?title=Chugach_Mountains *Contributors*: AlaskaTrekker, Backspace, Beeblebrox, BlindGoofy, Bryan Derksen, Calliopejen1, Decumanus, Dralwik, Gongoozler123, Hike395, James Crippen, Lafindumonde, Maxbeer, Postdlf, Quadell, RedWolf, Skookum1, Stan Shebs, Sten, Topbanana, Vedrfolnir, Zero Gravitas, 14 anonymous edits

Alaska Range *Source*: http://en.wikipedia.org/w/index.php?title=Alaska_Range *Contributors*: 1brownsfan, A little insignificant, AKMask, Abdulc, Acalamari, Acroterion, Alansohn, Alib, Andres, Astronautics, Bars 23, Beeblebrox, Bill Oaf, Bryan Derksen, CanisRufus, Click23, Dr. Blofeld, ESkog, Enviroboy, Foshizzlenizzlehoe, Gorotdi, Gotham77, Hike395, Jimmyptubas, Jonathan.s.kt, Librsh, Marshman, Mattisse, Maurreen, Merovingian, Possum, RedWolf, Romandial, Salamurai, Shannon1, Skookum1, Spireguy, Stan Shebs, Supra guy, The Thing That Should Not Be, Volcanoguy, Vsmith, Wikipe-tan, Wikiuser100, Xurei, Zero Gravitas, 50 anonymous edits

Alaska *Source*: http://en.wikipedia.org/w/index.php?title=Alaska *Contributors*: -Kerplunk-, 0, 12345llama, 1Martin33, 21655, 247balla34, 28421u2232nfenfcenc, 2busy2chat, 2tuntony, 5 albert square, 5150pacer, 732SOUTHPAW, 777fortytwo, A Sunshade Lust, A. B., A.K.A.47, A2Kafir, A5O123, A7x, ABCD, ABF, AKGhetto, ARUenergy, AVM, Aaron Einstein, Aarredondo6, Abce2, Abdul Muhib, Abdulc, Abramspoint, Abrech, Acanon, Aceia Express, Ackerman elementry, Acroterion, Adashiel, Addshore, AdjustShift, Adun12, Ahassan05, Ahernlj, Ahoerstemeier, Aitias, Aivazovsky, Ak kali, Akbeancounter, Akbikerpoet, Akdamien, Akgrl95, Akloki, Akwdb, Akxcskier, Alai, Alansohn, Alaska+yukon, AlaskaCruising, AlaskaMining, AlaskaTrekker, Alaskacatalog, Alaxsxa, Ale jrb, Aleksandr Grigoryev, Alethiareg, Alex Bakharev, Alex S, Alexagurl, AlexiusHoratius, Alexwcovington, Alfbar1, Algebraist, Ali'i, Alienbrett, AlistairMcMillan, Allstarecho, Alpha 4615, Alpodog999, Altenmann, Alxrnz2, Alyeska, Amaysky, Amazonien, American2, Amir beckham, AncientToaster, Andonic, Andre Engels, Andrewlp1991, Andrewmc123, Andrewpmk, Andy Christ, Andy Marchbanks, Andy120290, Andycjp, Anetode, Angbrown, Angela, Anger22, Angr, Animum, Anislhr, Anonymous Dissident, Antandrus, Anthony, Anthony22, AnthonyNgo, Antodav2007, Arakunem, Arbitrarily0, ArchonMagnus, ArielGold, Aris Katsaris, Arman88, Arnon Chaffin, Arsenikk, Art LaPella, Asdfgh, Ashanda, Astral, Astynax, Atanamir, Athene cunicularia, AuburnPilot, AuburnPilot, Aude, Aulis Eskola, Avenue, Axel23, AySz88, AzaToth, Azov, Aztom2, B. Fairbairn, BGManofID, BIL, BSveen, Backslash Forwardslash, Backspace, Bakamono, BankiSun, BanyanTree, Barbie xDoll, Bardeep18, Barek, Barrel-rider, Baseball Watcher, Basser g, Bazj, Bazzargh, Bearcat, Bearian, Beatlesfan27, Beeblebrox, Beland, BenBaker, Benjh40, Benscripps, Bento00, Betacommand, Bevo, Bibliomaniac15, Big Adamsky, Bigballdaddy, Bigkoiguy, Billgriswald, Bkamoua, Bkell, Bkonrad, Black-Velvet, Bladegirl, Blahstickman, Blalalalalaluaisi, Elanchardb, Bletch, Blobglob, Bluedogtn, Blueelectricstorm, Bmax10, Bob98133, BobJones, Bobblewik, Bobcat1122, BobertLePirate, Bobisbobbob, Bobo192, Bogey97, Bogie555, Bonacea, Bongwarrior, Bookandcoffee, Boone23howard, Borisborf, Bourquie, Boywiz, BradBeattie, Braden curci, Brainbark, BrendelSignature, Brewcrewer, BrianGV, Brianga, Brimba, Brion VIBBER, Broadcaster101, Brodens, Brodrocks1, Brucelee, BrunoBaIrd, Bryan Derksen, Brynath, Bswee, Btcisgod, Buaidh, Bubbabags, Bubliiz, Bullshark44, Bullzeye, Bunco man, Bunny Ann, Bunnyhop11, Burnordie4, Burntsauce, Butwhatdoiknow, C545, C777, CJLL Wright, CO, CPacker, CRKingston, CRON, CStyle, CWesling, CWii, Cakechild, Calaf, CalebNoble, Calliopejen1, Caltas, CambridgeBayWeather, Camerong, Can't sleep, clown will eat me, Can'tStandYa, CanadianLinuxUser, Canderson7, CanisRufus, Canuckian89, CapPixel, CapitalR, Caponer, Capricorn42, Captain Thor, Captain-tucker, Caputo737, CardinalDan, Carewser, Carl$Seemus, Carstonjacobsen, Cate, Catgut, Catlover1997, Causa sui, Cbrown1023, Cburnett, Ccacsmss, Cchow2, Cecil722, Celarnor, Celithemis, Cellin8r, Cenarium, Censusdata, Cg-realms, Chamal N, Chamberlian, Chandlerlink.04, Charles Matthews, CharlotteWebb, Chase me ladies, I'm the Cavalry, Chasingsol, Cherubino, ChessA4, Chico9955555, Chinook70, Chismisaballa, Chiwara, Chochopk, Chris 73, Chris G, Chris the speller, ChrisHodgesUK, ChrisJBenson, ChrisRuvolo, Chrisch, Chrisfow, Chrishmt0423, Chrislk02, Christian.B, Christy747, Chun-hian, Chunky Rice, Ciatochase, Cikoykip, Civil Engineer III, Cjs2111, Cjs56, Ckatz, Clamjam22, Clarince63, Clarkbhm, Claymore97, ClockworkSoul, Closedmouth, Clown57, Cluth, Cmano13, Cmichael, CodyBerryAKbadass, Coffee, Coffeeshivers, ColinMcMillen, Comboguard, Cometstyles, Commander Sergei Bjarkhov, CommonsDelinker, Communityprofiles, Condem, Conk 9, ConstantinetheGreat, Conversion script, Cool Stuff Is Cool, Cool3, Coolelyx, CooperDB, CopperSquare, CordeliaNaismith, Corinne68, Corriebertus, Corvus cornix, Cos111, Cottonshirt, Courcelles, Coyote sprit, CrazyC83, Creation7689, Creidieki, Cribananda, Crotalus horridus, Crowsnest, Csba, Ctjf83, Cubiq, Curps, Cyber topac, Cyktsui, CylonCAG, Cypher z, Cyrius, D Namtar, D0762, DARTH SIDIOUS 2, DCEdwards1966, DJ Clayworth, DMCer, DMacks, DOMICH, DSRH, DVD R W, Dabomb87, Dadaist6174, Dale Arnett, Damirgraffiti, Damyen185, DanMS, DancingPenguin, Danh, Daniel, Daniel,levine, Daniel5127, Dankirschner81, Danny, Daondo, Dark Mage, Dark Serge, Dark jedi requiem, Darolew, DarthVader, DasallmächtigeJ, DatVillain83, Dave.Dunford, Daven200520, Daverocks, Davewild, David Wahler, Davidmartell, Davidpinieiro, Dawn Bard, Daz 90, Dbrown123, Dcflyer, Dcornwall, DePortau, DeadEyeArrow, Deanshan, DebateLord, Debresser, Decumanus, Deevrod, Deflagro, Deflective, Deirdre, Dekimasu, Dekisugi, Delirium, Denali134, Denelson83, Dennis Bratland, Deor, DerHexer, DesmondRavenstone, DetroitHockey, Devin Pederson, DevinCook, Dewelar, Dextersexy the ripper, Dhtrible, Diannaa, Dicklyon, Digitalme, Dillard421, Diocles, Dirkbb, Discospinster, Dismas, DivineIntervention, Djsasso, Dkreisst, Dmh, Dmw, Dniete97, Doc glasgow, DocWatson42, Doctorbrassstone, Docu, Doczilla, DonKofAK, Donlammers, Doodmaster, DoubleBlue, Dougofborg, Doulos Christos, Dpaperman, Dposnow, Dpr, Dr Aaij, Dr. Blofeld, Dr.alf, Dr.finkelstein09, Dralwik, Droppedballz69, Drumguy8800, Drumlineramos, Duck that quacks alot, Dufekin, Dundana, Dust Filter, Dweens45, Dwim, Dybdal, Dyscti, Dysepsion, Dysmorodrepanis, Dysprosia, Długosz, E Pluribus Anthony, E2eamon, EGroup, EJF, ESkog, EaglesFanInTampa, Earldelawarr, Ebyabe, Eclecticology, Eco84, Ed Poor, Ed g2s, EdBever, Edivorce, Edward, Edwy, Egmontaz, EhsanQ, Ekneeley, Eknudson, El C, Elconejo, EldKatt, Electricalaskan, Elgreggo11, Elindros, Eliyahu S, EllenFoster, Elliotreed, Elliskev, Elsecar, Emax, Emil.B, Emuchick, En.ianm1121, EncycloPetey, Engelmann15, Enigmaman, Ennerk, Ent, Entrust101, Enviroboy, Eoghan888, Epbr123, Epicadam, Epolk, Eric Wester, Eric outdoors, Ericlaw02, Escape Orbit, Est.r, Esuzu, EugeneZelenko, Evb-wiki, Evertype, Everyking, Evice, Evil Monkey, Ewikdjmco, Excirial, Extreme outdoors, Eyekendra, FCYTravis, FF2010, Fabartus, Faithlessthewonderboy, Fang Aili, Faradayplank, Farsnickle, FastLizard4, Fastily, Fatty12345, Feedmecereal, Feitclub, Fenevad, Fernirm, Feyer, Fieldday-sunday, Fingerz, Finlay McWalter, Firetrap9254, First Light, Fisenko, Fishing, Fivestrokes, Fjehoel, Flatterworld, Flewis, FloK, Flockmeal, Flux.books, Flyguy649, FocalPoint, Fogherty V. Tatin, Footballfan190, Forrestleo, Foryst, Fossett&Elvis, Fqrstuvwxyz, Fractions, Frank Vest, Frankie0607, Fred Bradstadt, Fredbauder, Fredy.00, Free2conform123, FreplySpang, Frietjes, FriscoKnight, FritzG, Fritzpoll, Fry1989, Fsboalaska, Funkalunatic, Funnyfarmofdoom, Funnyhat, FusilEjerry86, Fvw, Fyyer, GDonato, Gadfium, Gaelen S., Gaff, Gail, Gaius Cornelius, Galloramenu, Gamingexpert, Garion96, Garravogue, Gary King, Gary cumberland, Gauss, Gbalaji82, Gblaz, Gcapp1959, Gebstadter, Geckoman1011, Geekgoddess24, Geekosaurus, GeoGreg, Geologyguy, George Burgess, George The Dragon, GerojiYuga, Gerrish, Gfoley4, Ghimboueils, Gholton, GhostPirate, Ghostreveries, Gia1156, Gilliam, Gilo1969, Gimmetrow, Gispert4, Giveostmoney, Glen, Glendoremus, Glubfig,

Gman2337, Gman411, Gmcfoley, Gnowor, Gogo Dodo, Golbez, GoldRingChip, Gonzo fan2007, Good Olfactory, GoodDay, Gordini53, Gotyear, Goumbaguy, Goustien, GraemeMcRae, Grafen, Graham87, GrahamColm, GrandfatherJoe, Grango242, Granola=D, Grapetonix, GregU, Gregorytruman, Gregsap, Grillo, Grim Revenant, Gritironskillet, Ground Zero, Grover cleveland, GrumpyTroll, Grunt, Gt, Guanaco, Gunmetal Angel, Gurch, Gurubrahma, HADC10, Haakon, Hadal, HaeB, HairyPerry, HalfShadow, Hammersbach, Hammersfan, Hammersoft, HangingCurve, Hanniganbaiter, Hapiee12, Happyfatman021, Happyharry345, Hardscarf, Harej, Harmil, Haruo, Hatter87, Hchrishicks, Hdt83, Heartinsanfrancisco, Heartlander, Heegoop, HeidiCandi, Hellnoha, Herdofhorses2005, HexaChord, Hfastedge, Hgav, Hibobjoe, Highfields, Highvale, Historychomper, HkCaGu, Hmains, Hmoul, Hobartimus, Hon-3s-T, Hookemdown, Hottentot, Howcheng, Hsfkwsf, g, Hughey, Huhsunqu, Hullaballoo Wolfowitz, Hungrymama, Husond, Hut 8.5, Hydrogen Iodide, I Own U284, IBook of the Revolution, II MusLiM HyBRiD II, IRP, IUnknown, IW.HG, Ianchreis, Ibrahmin, Icemanofbarcelona101, Igoldste, Ijackson2013, Ikariam3944, Immunize, Impulsion, Imyaabaian, In Defense of the Artist, In fact, Indon, Infoporfin, Inhumer, Instinct, Inter, Inwind, Ioij, Irnavash, Irregulargalaxies, Isomage, Itai, Itscolt, Ivangrozny69, Ixfd64, Izehar, J Crow, J Di, J Readings, J. Daily, J.delanoy, JCRB, JD554, JDJintheAK, JEdmundson, JForget, JMK, JNW, Ja 62, JaGa, JacobS, JadziaLover, Jaguar, Jak123, Jake Zhang, Jakew, Jakewater, Jaksmata, James Crippen, JamesMLane, JamesReyes, Jamesooders, Jamisonhalliwell, Java7837, Jay Litman, Jay Zaq, Jayron32, Jbamb, Jcam, Jcarman, Jclemens, Jcmiller1215, Jct alaska, JediScougale, Jeff G., JeffPGibson, Jefferythegreat, Jeffmedkeff, JeffreyAllen1975, Jelinek121, Jellotine, Jeltz, Jemmy Button, Jengod, Jeremiestrother, Jeremyhen, Jeronimo, Jerry5916, JerryVanF, Jerrycallo, Jesikahilton, Jesse Viviano, JetLover, Jfisto4life, Jh51681, Jhartmann, Jhendin, JimIrwin, JimWae, Jimp, JinJian, Jinian, Jiy, Jj137, Jkaufman101, Jlaramee, Jlittlet, Joelr31, John K, John Reaves, John Vandenberg, JohnInDC, JohnJHenderson, Johnman239, Jojhutton, Jonassweden, Jonathan Kovaciny, Jonathan.s.kt, Jonathunder, Jonemerson, Jonndk, Jonwillig, Josborne2382, Jose5000, Jose77, Joseph Solis in Australia, Josephhartman07, Joshua Issac, JossBuckle Swami, Jtkiefer, Juan Cruz, Juantay, Judeeclare, Juliancolton, Jumba lumba, Jusdafax, Jusjih, Just James, Justinbieberizsmexi, Justinmcl, Juzeris, K. Annoyomous, K50 Dude, KCinDC, KFP, KGV, KJS77, KQW, Kaare, Kablammo, Kabri, Kafziel, Kahuroa, Kairos, Karam.Anthony.K, KarasuGamma, Kareeser, Karl6109, Katalaveno, KathrynLybarger, Kavas, Kazrak, Kbdank71, Kbthompson, Keelm, Keilana, KeithH, Kelly, Ken Gallager, Ken g6, Keptarkoto, Kerotan, Ketchikanadian, Kether83, KevinR, KevinSun2000, KevinTR, Kevincuth08, Kewldude246810, Killer monkey 95, Killervogel5, Killiondude, King of Hearts, Kingal86, Kingpin13, Kintetsubuffalo, Kirachinmoku, Kitch, Kkkiii, Klaustronaut, KnightLago, KnowledgeOfSelf, Knutux, KokkaShinto, Konman72, Kookykman, Koolkidz123, Kotjze, Kowh, Kozuch, Krashlandon, Krellis, Kross, Kryptonian250, Kungfuadam, Kungming2, Kuo.mintang, Kusma, Kuzain, Kwamikagami, Kwertii, L'Aquatique, LEDfarmer, LGagnon, LOL, La Pianista, La goutte de pluie, LafinJack, Lan56, Lasttooth, Latitude0116, Laughonthefloor, Laurascudder, Lavalette1, LedgendGamer, LeheckaG, Lenticel, Levineps, LibLord, Lightmouse, Lil shaf, Lilac Soul, Lilpinoy 82, Ling.Nut, LinnySawrus, LizardJr8, Lk9, Llamadog903, Londonbats, Longhairededed, Lord Voldemort, Lotje, Lougire, Lowell33, Lrdwhyt, Luckylettuce, Luk, LukeHogg456, Luna Santin, Luong, Luvinmiley4ever, Lysior, Lzf, M C Y 1008, MBisanz, MER-C, MJ94, MJCdetroit, MLHarris, MONGO, MPF, MPerel, MYHelper, MackSalmon, Madchester, Madhero88, Madmedea, Magioladitis, Magister Mathematicae, Maias, Mak121, Makaristos, Malcolm Farmer, Mananaliksik, Mandarax, Manderiko, Mani1, Manjithkaini, Marek69, Markaci, Markjmeyer7, Marqueed, Marquisjamesphelan, Martarius, Martin S Taylor, Martin451, Martinwilke1980, Mastrchf91, Matchups, Matijap, Mato, Matt skrlac, Matt.T, MattWright, Mattbrundage, Mattisse, Mav, Maxis ftw, Mcnastybllr69, Mcpusc, Meaghan, Megaboz, Megamouthbolt, Melsaran, Member, Mentisock, Mercer5089, Mercutio84110, Merovingian, Mesmes, Metron4, Mgdowney, Mic, MichaelSpeer, Michaelas10, Michaelbusch, Michaeldsuarez, Mifter, Mightymights, Mike Rosoft, Mike s, MikeJa2, Mikeblas, Mikemike12, Mikevegas40, Mindmatrix, MindstormsKid, Minesweeper, Minimac's Clone, Minna Sora no Shita, Miskwito, Missouri Class, Mistakenjam, Mister me88, Mithridates, Miyokan, Mobile Snail, Modelmany, Moe Epsilon, Mohonu, Moncrief, Moneybag990, Monkey Bounce, Monkeyman, Monkeymania302, Monobi, Montana's Defender, Monterey Bay, Montrealais, Monty845, Mooore, Mooshiga, Moscownews, Moulder, Moverton, Moxy, Mr.Sexman, Mr.crabby, MrD9, MrMacMan, Mrbigg9969, Mrgagafoo, Mrjinx, Mrstraty32, Mscuthbert, Msp0, Mu Cow, MuZemike, Muj0, MuscleJaw SobSki, Mushroom, Muzzy12345678987654321, Mwanner, Mx3, Mxn, Mygerardromance, Mykej, Mynameisbobby2, Myrtlecharlotte, Mysdaao, Mzajac, N734LQ, NCurse, NEICenergy, NHRHS2010, Nakon, Nancy, Narutolovehinata5, Nascar1996, Nat Krause, Natl1, Navamske27, NawlinWiki, Nburden, NeilN, Neitherday, Neofelis Nebulosa, NetBMC, Nethency, Netoholic, Netsnipe, Neutrality, Neverquick, Newsaholic, Newshound08, Nhelm83, NicAgent, Nickgonzalez, Nickhuh, Niel Orcutt, Nightkey, Ninja kei, Nishkid64, Nixeagle, Nkrosse, Nlu, Nneonneo, No Guru, Noll123, Non't deny my share., Noobtube3, Northwesterner1, NotACow, Notbyworks, Notscott, Nouse4aname, Nowayjose1221, Nparra, Nsaa, Ntse, Nubiatech, NuclearWarfare, Nukeless, Nuttycoconut, Nwbeeson, Ny156uk, Nyttend, Nyyr2cool1, Oakshade, Oanabay04, Oda Mari, Oddnessly, Odie5533, Off the heezay, Ogno, Ohconfucius, Ohnoitsjamie, Ojigiri, Olive wombat, Olivier, Omdo, Omeomi, Omgee, Omnieiunium, Ondewelle, OnePt618, Onthegogo, Opelio, Optimist on the run, Orcalover, Ortcutt, Ottokarten, Ouro, Outlando, Overmaster net, Ovis23, Oxymoron83, Pabouk, Pack3406, Packers3789, Paine, Paine Ellsworth, Pan Dan, Pandamoania, Patrick, Pattond2, Paul August, Paul Erik, Pavelegorov, Pax:Vobiscum, PeaceNT, Peaceduck, Pedroacosta345, Peli1414, Penwhale, Perey, Perfect Proposal, Peridon, Perkyville, Persian Poet Gal, Peruvianllama, Peter Horn, Petersam, Petter Strandmark, Peyre, Pfly, Pgk, Pglitsch, Pharaoh of the Wizards, Philg88, Philip Trueman, Philosofool, Phizzy, Phlarsen, Phoenix79, Phydend, Piano non troppo, Piast93, Pigman, Piledhigheranddeeper, Pilotguy, Pinethicket, Pip2andahalf, Pisces116, Pjrich, Plastikspork, Pmac5, Pmresch, Pmt7ar, Poindexter Propellerhead, PollShark, Polly, PookeyMaster, Poopdeck90210, Porchwhen, Porlock6, Porqin, Postdlf, Pr1268, Pras, Premiercolleges, Presidentialness, Presidentluis, Prestonmag, Processed meat, Prodego, Professor Davies, Professor marginalia, Promethean, Proski, Psantora, Ptcamn, Pzavon, Qaddosh, Qqqqqq, Qt ak, Quadell, Quadrius, Quantpole, QueenCake, Queson, Quest for Truth, Quinn00, Quintote, Quizimodo, Quota, Qutezuce, Quuxplusone, Qwertasdfgh, Qxz, R'n'B, R9tgokunks, RA0808, RabinK007, Racepacket, RadioKAOS, Radioshed, Radon210, Rahzel, Raichu, RainbowOfLight, Ram-Man, Ramurf, Random moi, Random user 39849958, Randomguythatsbored, RandorXeus, Rapthorne, Ravedave, Raven097, Rawrkitty099, Ray Van De Walker, Raymond, Razorflame, Realestatewiki, Realm of Shadows, Realm of the crimson viper, Reaper Eternal, Rearden Metal, Receptacle, Reconfirmer, Redrocket, Redthoreau, Reesh, Reinyday, Restre419, Retired username, Rettetast, Reuvenk, Revas, RexNL, Reywas92, Rfc1394, Rhatsa26X, Rholton, Riana, Rich Farmbrough, Rich257, Richard David Ramsey, Richard W.M. Jones, Richardcavell, Richi, Rick Block, Ricky81682, RiverFattieRCool, Rjd0060, Rje, Rjjo, Rjm656s, Rjwilmsi, Rkavuru, Rkstafford, Rob Hooft, RobLa, RobertG, RobertGustafson, Roche-Kerr, Rock nj, Rocketmaniac, Rockonjake4996, RockyMtnGuy, Roger Pilgham, Roke, Romandial, Romanm, Ron Ritzman, Ronebofh, Rorschach, Rossdegenstein, Rossumcapek, RoyBoy, Rrburke, Rreagan007, Rrius, Rror, Rs09985, Ruhrfisch, Rumping, Rupertslander, Russianman12, Ruth-2013, Rwk, Ryulong, SDC, SEWilco, SGT141, SHallathome, SPUI, Saberwolf116, Sackadatfunk, Sadads, Saebhiar, SaltyBoatr, Saluyot, Sam Hocevar, Sam Korn, Sambrougher, Samuel 69105, SandyGeorgia, Sango123, Saopaulo1, Saravask, Sart91, SatyrTN, SaulPerdomo, Savant13, Saywhatzannuuhh, Sburke, ScaldingHotSoup, Scanadiense, Scanlan, Scarab12, Scarian, Sceptre, Schcambo, SchuminWeb, Scientizzle, Sciurinæ, Scm83x, Scott Burley, Scott Illini, ScottMainwaring, ScottSteiner, Scoutersig, Scriberius, Scythian1, Seansinc, Seaphoto, Seb az86556, Sedna10387, Seegoon, Seglea, Seidenstud, Selxxa, Semperf, Senrable, Seraphim, Sergeantkitty, Sfan00 IMG, Sfmontyo, Sgeureka, Shadowjams, ShakespeareFan00, Shamir1, Shanel, Shanes, Shantavira, Shatar, Shawisland, Shawn alexander, Sheogorath, Shimgray, Shinjiman, Shirulashem, Shoeofdeath, Shoopza, Shorne, Shunpiker, Silington, Simon Lieschke, SimonP, Simondarren21, Sirgregmac, Sjakkalle, Sk8ski, Skarebo, Sketchmoose, Skew-t, SkinnerSix, Skookum1, Skshow, Sky Attacker, Skybum, Skywayman, Slakr, Slarson, Sleddoggin, Slicky, Sligocki, Slowking Man, Smallbones, SmartGuy, Smee, SmilesALot, Smith03, Smokizzy, SmthManly, Smumdax, Snowboarder, Snowmanradio, Snowolf, Soccerjpm13, Soccerplayer5, Soerenfm, Soetermans, Solipsist, Somarinoa, Some jerk on the Internet, Sonia, SonicAD, Soobrickay, Sophos II, SorryGuy, SoulMeetsBody, South Bay, SouthernMan, SpaceRocket, Spartan-James, Spawn Man, Speakandspell, SpecMode, Spike Wilbury, Spliffy, SpookyMulder, Sting-fr, Student7, Studerby, Studmuffinsupreme2, Subtitledryhump, Suffusion of Yellow, Sunray, SupaStarGirl, Super-Magician, SuperHamster, Superbeecat, Superpirate, Supspirit, Sus scrofa, Susvolans, Svick, Swimm21, Swollib, Swpb, Sylent, Synchronism, Szajci, T.J.V., THEN WHO WAS PHONE?, TOttenville8, TShilo12, TUF-KAT, TaerkastUA, Taggard, Taikoguy36, TakisV, Tanker58er, Taranah, Tarkan1st, Tassedethe, Taylorbio, Tbennert, Tbhotch, Tcoahran, Techman224, TedCroushore, Tedder, Tedrader, Teknomunk, Tellyaddict, Template namespace initialisation script, Tesscass, Tex, TexasAndroid, The Anome, The Blade of the Northern Lights, The Duke of Waltham, The Epopt, The Handshake compromise, The High Fin Sperm Whale, The Man in Question, The Obento Musubi, The Rambling Man, The Storm Surfer, The Thing That Should Not Be, The Transhumanist, The Universe Is Cool, The cheeseburglar, The sock that should not be, The sunder king, The wub, TheHYPO, TheNewPhobia, TheRanger, TheSpectator, Thecurran, Theda, Thedjb, Thegarland, Thegreatglobetrotter, Thehistoryprofessor, Thejamie024, Thelb4, Theman1313, Thesouthernhistorian45, Thetruththeliesthesecrets, Thief12, Thingg, ThinkBlue, Thirty-seven, Thivierr, Thorwald, Thundermist04167, Tiddly Tom, Tide rolls, Tidying Up, TigerShark, Tim Q. Wells, Tim Shell, Timberframe, Timmeh, Timneu22, TimothyHorrigan, Timwi, Tiptoety, Tiredthird, Tixity, Tkasmai, Tkynerd, Tm0ney3, Tmopkisn, Tobby72, Todd on a Tractor, Tomeasy, Tommy2010, Tonei, Tony Fox, Tony Stevens, Tony1, Toropop, Torterrafire, Tpbradbury, Tracer9999, Tre1914, Triggs, Triona, TriviaMan!, Triwbe, Trusilver, Tsange, Tsdek, Ttony21, Ttsalo, Tuggler, Turlo Lomon, Twig732, TwinCityIL, TwistOfCain, Two-face Jackie, Ufwuct, Ulric1313, Ultranotadork, Uncle Dick, Unity-north, Unschool, Uris, Ute in DC, VVPushkin, Valfontis, Vamrat, Vancouverovka, Vanished User 8a9b4725f8376, Vanjagenije, VasilievVV, Vaticrat, Vcelloho, Vectro, Vedrfolnir, VegaDark, Velvetron, Venice, Verrai, Versageek, Versus22, Vicki Rosenzweig, Vidor, Vikrant42, Vina, Vishnava, Voksen, Voskoboinikov, Voyagerfan5761, Vpuliva, Vrinan, Vsmith, Vuvar1, WDavis1911, Wafulz, Waggers, Wapcaplet, Wars, Watcharakorn, WaterMelon7, Wayne Slam, Wayne312, Wayward, Weeliljimmy, Weetoddid, Wenli, Weregerbil, Wes!, Wester, Whereizben, WhisperToMe, White 720, White Shadows, Whittsnake, Why Not A Duck, Wiff321, Wiki alf, Wiki0709, WikiDao, Wikien2009, WikipedianMarlith, Wikipelli, Wikispork, Wikster72, Wild Wolf, Wilee, Will Beback, Will Beback Auto, William Avery, WilliamH, Willking1979, Wireless Keyboard, Witchwooder, Wkeith112003, Wknight94, Wmahan, Wmurray4452, Wolfmantheo, Woodstein52, Woohookitty, Woolters, Woscafrench, Wukkuan, Wvelite20, Wwoods, Xenghornt, Xepher, Xezbeth, Xiahou, Xiaoyu of Yuxi, Xiner, Xompanthy, Xorkl000, Xp54321, Xyzzyva, Yamamoto Ichiro, Yankee Rajput, Yarnalgo, Ye Olde Luke, YeshuaDavid, YixilTesiphon, Yksin, Ylee, Yms, Yopienso, Yourbestfriendisblack, Youssefsan, Yrodro, Ysangkok, Yupik, Yutsi, Zabadinho, Zaian, Zakkuro, Zaledin, ZapThunderstrike, Zarex, Zazaban, Zeimusu, Zel95, Zephyris, Zero Gravitas, Zerotran0, Zewu, Ziansh, Zigger, Zigzug, Zoe, Zsinj, Zyxw, Zzuuzz, Zzyzx11, Zzyzx513, ^demon, Δ, 'Бꙑль, , 3782 anonymous edits

Image Sources, Licenses and Contributors

Image:Alaska 8 shield.svg *Source*: http://en.wikipedia.org/w/index.php?title=File:Alaska_8_shield.svg *License*: unknown *Contributors*: SPUI, Sevela.p

Image:Alaska 3 shield.svg *Source*: http://en.wikipedia.org/w/index.php?title=File:Alaska_3_shield.svg *License*: unknown *Contributors*: SPUI, Sevela.p

Image:Alaska 4 shield.svg *Source*: http://en.wikipedia.org/w/index.php?title=File:Alaska_4_shield.svg *License*: unknown *Contributors*: SPUI, Sevela.p

Image:Denali Highway.jpg *Source*: http://en.wikipedia.org/w/index.php?title=File:Denali_Highway.jpg *License*: unknown *Contributors*: User:JKBrooks85

Image:Picea glauca taiga.jpg *Source*: http://en.wikipedia.org/w/index.php?title=File:Picea_glauca_taiga.jpg *License*: unknown *Contributors*: L.B. Brubaker

File:Mcclarensummit.JPG *Source*: http://en.wikipedia.org/w/index.php?title=File:Mcclarensummit.JPG *License*: unknown *Contributors*: Beeblebrox (talk). Original uploader was Beeblebrox at en.wikipedia

File:Denaliesker.JPG *Source*: http://en.wikipedia.org/w/index.php?title=File:Denaliesker.JPG *License*: unknown *Contributors*: Beeblebrox at en.wikipedia

File:Denalifromdenali.JPG *Source*: http://en.wikipedia.org/w/index.php?title=File:Denalifromdenali.JPG *License*: unknown *Contributors*: Beeblebrox (talk). Original uploader was Beeblebrox at en.wikipedia

File:Lakesalaskarange.JPG *Source*: http://en.wikipedia.org/w/index.php?title=File:Lakesalaskarange.JPG *License*: unknown *Contributors*: Beeblebrox (talk). Original uploader was Beeblebrox at en.wikipedia

File:Susitnabridge.JPG *Source*: http://en.wikipedia.org/w/index.php?title=File:Susitnabridge.JPG *License*: unknown *Contributors*: Beeblebrox (talk). Original uploader was Beeblebrox at en.wikipedia

Image:Alaska 1 shield.svg *Source*: http://en.wikipedia.org/w/index.php?title=File:Alaska_1_shield.svg *License*: unknown *Contributors*: SPUI, Sevela.p

Image:I-A1.svg *Source*: http://en.wikipedia.org/w/index.php?title=File:I-A1.svg *License*: unknown *Contributors*: Aconcagua, Augiasstallputzer, Fietsbel, Ltljltlj, Raquel350, SPUI, 1 anonymous edits

Image:Alaska 2 shield.svg *Source*: http://en.wikipedia.org/w/index.php?title=File:Alaska_2_shield.svg *License*: unknown *Contributors*: Conscious, Denelson83, SPUI, Sevela.p

Image:I-A2.svg *Source*: http://en.wikipedia.org/w/index.php?title=File:I-A2.svg *License*: unknown *Contributors*: Aconcagua, Augiasstallputzer, Fietsbel, Ltljltlj, Raquel350, SPUI, 1 anonymous edits

File:1st car on Richardson Highway 1913.jpg *Source*: http://en.wikipedia.org/w/index.php?title=File:1st_car_on_Richardson_Highway_1913.jpg *License*: unknown *Contributors*: nps.gov

Image:Meadow lake along Richardson Highway, Alaska.jpg *Source*: http://en.wikipedia.org/w/index.php?title=File:Meadow_lake_along_Richardson_Highway,_Alaska.jpg *License*: unknown *Contributors*: Delirium, Korg

File:Richardson Highway Badger Road Interchange.jpg *Source*: http://en.wikipedia.org/w/index.php?title=File:Richardson_Highway_Badger_Road_Interchange.jpg *License*: unknown *Contributors*: User:RadioKAOS

File:Heatpipes.JPG *Source*: http://en.wikipedia.org/w/index.php?title=File:Heatpipes.JPG *License*: unknown *Contributors*: User:Beeblebrox

Image:AK_map_Denali.svg *Source*: http://en.wikipedia.org/w/index.php?title=File:AK_map_Denali.svg *License*: unknown *Contributors*: Cobaltcigs

Image:Mount McKinley Alaska.jpg *Source*: http://en.wikipedia.org/w/index.php?title=File:Mount_McKinley_Alaska.jpg *License*: unknown *Contributors*: Nic McPhee from Morris, MN, USA

Image:Denali National Park.jpg *Source*: http://en.wikipedia.org/w/index.php?title=File:Denali_National_Park.jpg *License*: unknown *Contributors*: Jennifer from Wichita, USA

Image:Grizzly Denali edit.jpg *Source*: http://en.wikipedia.org/w/index.php?title=File:Grizzly_Denali_edit.jpg *License*: unknown *Contributors*: User:Fir0002

File:Heinrich Berann NPS Denali.jpg *Source*: http://en.wikipedia.org/w/index.php?title=File:Heinrich_Berann_NPS_Denali.jpg *License*: unknown *Contributors*: Heinrich Berann

File:Denali-alpine-lakes-forest-Highsmith.jpeg *Source*: http://en.wikipedia.org/w/index.php?title=File:Denali-alpine-lakes-forest-Highsmith.jpeg *License*: unknown *Contributors*: Carol M. Highsmith (1946–)

Image:DenaliNationalPark.JPG *Source*: http://en.wikipedia.org/w/index.php?title=File:DenaliNationalPark.JPG *License*: unknown *Contributors*: 127x0x0x1, Gay Cdn, Marianocecowski, 2 anonymous edits

File:Denali National Park Polychrome Mountains Panorama 27193px.jpg *Source*: http://en.wikipedia.org/w/index.php?title=File:Denali_National_Park_Polychrome_Mountains_Panorama_27193px.jpg *License*: unknown *Contributors*: Photo (c)2007 Derek Ramsey (Ram-Man)

File:Magnify-clip.png *Source*: http://en.wikipedia.org/w/index.php?title=File:Magnify-clip.png *License*: unknown *Contributors*: User:Erasoft24

Image:I-A4.svg *Source*: http://en.wikipedia.org/w/index.php?title=File:I-A4.svg *License*: unknown *Contributors*: Aconcagua, Augiasstallputzer, Fietsbel, Ltljltlj, Raquel350, SPUI, 1 anonymous edits

Image:ParksHighwayatHurricane.jpg *Source*: http://en.wikipedia.org/w/index.php?title=File:ParksHighwayatHurricane.jpg *License*: unknown *Contributors*: User:Christy747

File:Parks Highway to Fairbanks.jpg *Source*: http://en.wikipedia.org/w/index.php?title=File:Parks_Highway_to_Fairbanks.jpg *License*: unknown *Contributors*: Jim from Lexington, KY, USA

Image:Milepost238.jpg *Source*: http://en.wikipedia.org/w/index.php?title=File:Milepost238.jpg *License*: unknown *Contributors*: User:Christy747

File:Roundtanglelake.JPG *Source*: http://en.wikipedia.org/w/index.php?title=File:Roundtanglelake.JPG *License*: unknown *Contributors*: Beeblebrox (talk). Original uploader was Beeblebrox at en.wikipedia

File:Chitina dipnet.jpg *Source*: http://en.wikipedia.org/w/index.php?title=File:Chitina_dipnet.jpg *License*: unknown *Contributors*: Aconcagua, BanyanTree, Dankarl, Stunteltje

File:Miles Glacier Bridge, damage and kludge, 1984.jpg *Source*: http://en.wikipedia.org/w/index.php?title=File:Miles_Glacier_Bridge,_damage_and_kludge,_1984.jpg *License*: unknown *Contributors*: Choess, Closeapple, Denimadept, Dogears, Ronaldino, 1 anonymous edits

Image:Copper1.jpg *Source*: http://en.wikipedia.org/w/index.php?title=File:Copper1.jpg *License*: unknown *Contributors*: AlaskaTrekker

File:Copper River fishwheels.jpg *Source*: http://en.wikipedia.org/w/index.php?title=File:Copper_River_fishwheels.jpg *License*: unknown *Contributors*: Doug Noon

Image:Copper2.jpg *Source*: http://en.wikipedia.org/w/index.php?title=File:Copper2.jpg *License*: unknown *Contributors*: AlaskaTrekker

File:Copper River near Chitina.jpg *Source*: http://en.wikipedia.org/w/index.php?title=File:Copper_River_near_Chitina.jpg *License*: unknown *Contributors*: Doug Noon

Image:Picea mariana taiga.jpg *Source*: http://en.wikipedia.org/w/index.php?title=File:Picea_mariana_taiga.jpg *License*: unknown *Contributors*: Aconcagua, MPF, Zejo, 1 anonymous edits

Image:Glacial Dust off Alaska.jpg *Source*: http://en.wikipedia.org/w/index.php?title=File:Glacial_Dust_off_Alaska.jpg *License*: unknown *Contributors*: Jeff Schmaltz

Image: MountWrangell.jpg *Source*: http://en.wikipedia.org/w/index.php?title=File:MountWrangell.jpg *License*: unknown *Contributors*: Aconcagua, Rémih, Tolanor, Urban

Image: Wrangell Mountains.jpg *Source*: http://en.wikipedia.org/w/index.php?title=File:Wrangell_Mountains.jpg *License*: unknown *Contributors*: USGS

Image:Wrangells1.jpg *Source*: http://en.wikipedia.org/w/index.php?title=File:Wrangells1.jpg *License*: unknown *Contributors*: Original uploader was AlaskaTrekker at en.wikipedia

Image:Wrangells Mountains Alaska.jpg *Source*: http://en.wikipedia.org/w/index.php?title=File:Wrangells_Mountains_Alaska.jpg *License*: unknown *Contributors*: Aconcagua, Urban

Image: Chugach Panorama Alaska.jpg *Source*: http://en.wikipedia.org/w/index.php?title=File:Chugach_Panorama_Alaska.jpg *License*: unknown *Contributors*: Aconcagua, Urban

Image:Chugach4.jpg *Source*: http://en.wikipedia.org/w/index.php?title=File:Chugach4.jpg *License*: unknown *Contributors*: Original uploader was AlaskaTrekker at en.wikipedia

Image:Chugachreflection.JPG *Source*: http://en.wikipedia.org/w/index.php?title=File:Chugachreflection.JPG *License*: unknown *Contributors*: Beeblebrox (talk). Original uploader was Beeblebrox at en.wikipedia

Image: Peaks of the Alaska Range (1).jpg *Source*: http://en.wikipedia.org/w/index.php?title=File:Peaks_of_the_Alaska_Range_(1).jpg *License*: unknown *Contributors*: Frank K. from Anchorage, Alaska, USA

File: Alaska_range.jpg *Source*: http://en.wikipedia.org/w/index.php?title=File:Alaska_range.jpg *License*: unknown *Contributors*: User:1brownsfan

Image:Alaska Range Glacier.jpg *Source*: http://en.wikipedia.org/w/index.php?title=File:Alaska_Range_Glacier.jpg *License*: unknown *Contributors*: Jack French from San Francisco, USA

Image:MountMcKinley BA.jpg *Source*: http://en.wikipedia.org/w/index.php?title=File:MountMcKinley_BA.jpg *License*: unknown *Contributors*: Calliopejen1, Donut, Urban, 2 anonymous edits

File:Alaska Range Mountain Peaks.jpg *Source*: http://en.wikipedia.org/w/index.php?title=File:Alaska_Range_Mountain_Peaks.jpg *License*: unknown *Contributors*: U.S. Fish and Wildlife Service. Original uploader was Mattisse at en.wikipedia

File:Craggyakrange.JPG *Source*: http://en.wikipedia.org/w/index.php?title=File:Craggyakrange.JPG *License*: unknown *Contributors*: User:Beeblebrox

File:Gulkanaglacier.JPG *Source*: http://en.wikipedia.org/w/index.php?title=File:Gulkanaglacier.JPG *License*: unknown *Contributors*: User:Beeblebrox

File:Flag of Alaska.svg *Source*: http://en.wikipedia.org/w/index.php?title=File:Flag_of_Alaska.svg *License*: unknown *Contributors*: Anime Addict AA, AnonMoos, Dbenbenn, DevinCook, Duck that quacks alot, Dzordzm, Fry1989, Homo lupus, Juiced lemon, Juliancolton, MGA73, Mattes, Nightstallion, R2D2Art2005, Resident Mario, Serinde, Smooth O, VIGNERON, Wester, Zscout370, 10 anonymous edits

File:Alaska-StateSeal.svg *Source*: http://en.wikipedia.org/w/index.php?title=File:Alaska-StateSeal.svg *License*: unknown *Contributors*: U.S. Government

File:Map of USA AK full.svg *Source*: http://en.wikipedia.org/w/index.php?title=File:Map_of_USA_AK_full.svg *License*: unknown *Contributors*: User:Skew-t

Image:Speakerlink.svg *Source*: http://en.wikipedia.org/w/index.php?title=File:Speakerlink.svg *License*: unknown *Contributors*: Woodstone. Original uploader was Woodstone at en.wikipedia

File:Alaska area compared to conterminous US.svg *Source*: http://en.wikipedia.org/w/index.php?title=File:Alaska_area_compared_to_conterminous_US.svg *License*: unknown *Contributors*: User:Sting

File:Denali Mt McKinley.jpg *Source*: http://en.wikipedia.org/w/index.php?title=File:Denali_Mt_McKinley.jpg *License*: unknown *Contributors*: RedWolf

File:Grizzly Bear Fishing Brooks Falls.jpg *Source*: http://en.wikipedia.org/w/index.php?title=File:Grizzly_Bear_Fishing_Brooks_Falls.jpg *License*: unknown *Contributors*: User:Azov

File:Augustine Volcano Jan 12 2006 edited-1.jpg *Source*: http://en.wikipedia.org/w/index.php?title=File:Augustine_Volcano_Jan_12_2006_edited-1.jpg *License*: unknown *Contributors*: Game McGimsey

File:Public-Lands-Western-US.png *Source*: http://en.wikipedia.org/w/index.php?title=File:Public-Lands-Western-US.png *License*: unknown *Contributors*: User:Northwest-historian

File:Barrow beach.jpg *Source*: http://en.wikipedia.org/w/index.php?title=File:Barrow_beach.jpg *License*: unknown *Contributors*: Calliopejen, Cromagnon, Urban

File:AlutiiqDancer.jpg *Source*: http://en.wikipedia.org/w/index.php?title=File:AlutiiqDancer.jpg *License*: unknown *Contributors*: Christopher Mertl

File:Russian Sloop-of-War Neva.jpg *Source*: http://en.wikipedia.org/w/index.php?title=File:Russian_Sloop-of-War_Neva.jpg *License*: unknown *Contributors*: Drawn by Capt Lisiansky, engraved by I. Clark. Published by John Booth, Duke Street, Portland Place, London, 1 March 1814

File:Miners climb Chilkoot.jpg *Source*: http://en.wikipedia.org/w/index.php?title=File:Miners_climb_Chilkoot.jpg *License*: unknown *Contributors*: Hegg, E.A (1867-1948)

File:AttuSnow.jpg *Source*: http://en.wikipedia.org/w/index.php?title=File:AttuSnow.jpg *License*: unknown *Contributors*: AustralianRupert, Cla68, Cobatfor, Docu, Prüm

File:Russian Orthodox Church.jpg *Source*: http://en.wikipedia.org/w/index.php?title=File:Russian_Orthodox_Church.jpg *License*: unknown *Contributors*: Barek, Calliopejen, Ebyabe, FlickrLickr, FlickreviewR, Jmabel, Juiced lemon, Kurpfalzbilder.de, PereslavlFoto, Wst, 2 anonymous edits

File:Prudhoe Bay aerial FWS.jpg *Source*: http://en.wikipedia.org/w/index.php?title=File:Prudhoe_Bay_aerial_FWS.jpg *License*: unknown *Contributors*: U.S. Fish and Wildlife Service

File:Bpanchorage.JPG *Source*: http://en.wikipedia.org/w/index.php?title=File:Bpanchorage.JPG *License*: unknown *Contributors*: User:Beeblebrox

File:Alaska Airlines, N767AS.jpg *Source*: http://en.wikipedia.org/w/index.php?title=File:Alaska_Airlines,_N767AS.jpg *License*: unknown *Contributors*: Frank K. from Anchorage, Alaska, USA

File:Fairbanks Memorial Hospital.jpg *Source*: http://en.wikipedia.org/w/index.php?title=File:Fairbanks_Memorial_Hospital.jpg *License*: unknown *Contributors*: User:RadioKAOS

File:Alyeska Resort.JPG *Source*: http://en.wikipedia.org/w/index.php?title=File:Alyeska_Resort.JPG *License*: unknown *Contributors*: Eric Goedtel

File:Alaska Pipeline Closeup Underneath.jpg *Source*: http://en.wikipedia.org/w/index.php?title=File:Alaska_Pipeline_Closeup_Underneath.jpg *License*: unknown *Contributors*: Photo by and (c)2005 Derek Ramsey (Ram-Man)

File:Alaska Crude Oil Reserves.PNG *Source*: http://en.wikipedia.org/w/index.php?title=File:Alaska_Crude_Oil_Reserves.PNG *License*: unknown *Contributors*: User:RockyMtnGuy

File:Alaska Crude Oil Production.PNG *Source*: http://en.wikipedia.org/w/index.php?title=File:Alaska_Crude_Oil_Production.PNG *License*: unknown *Contributors*: User:RockyMtnGuy

File:Pacific Halibut Fileting.JPG *Source*: http://en.wikipedia.org/w/index.php?title=File:Pacific_Halibut_Fileting.JPG *License*: unknown *Contributors*: Jlikes2Fish

File:Seward Highway.jpg *Source*: http://en.wikipedia.org/w/index.php?title=File:Seward_Highway.jpg *License*: unknown *Contributors*: steve lyon from los angeles, ca, usa

File:AlaskaRailroad.jpg *Source*: http://en.wikipedia.org/w/index.php?title=File:AlaskaRailroad.jpg *License*: unknown *Contributors*: Christy747 Original uploader was Christy747 at en.wikipedia

File:Tustumena, Alaska Marine Highway.jpg *Source*: http://en.wikipedia.org/w/index.php?title=File:Tustumena,_Alaska_Marine_Highway.jpg *License*: unknown *Contributors*: User:NancyHeise

File:Alaska-737-4QB-YVR.jpg *Source*: http://en.wikipedia.org/w/index.php?title=File:Alaska-737-4QB-YVR.jpg *License*: unknown *Contributors*: User:Makaristos

File:Juneau, Alaska Downtown.jpg *Source*: http://en.wikipedia.org/w/index.php?title=File:Juneau,_Alaska_Downtown.jpg *License*: unknown *Contributors*: pdx3525

File:Anchorage1.jpg *Source*: http://en.wikipedia.org/w/index.php?title=File:Anchorage1.jpg *License*: unknown *Contributors*: Aconcagua, Xnatedawgx

File:Fairbanks05.jpg *Source*: http://en.wikipedia.org/w/index.php?title=File:Fairbanks05.jpg *License*: unknown *Contributors*: Original uploader was JeffreyAllen1975 at en.wikipedia Later version(s) were uploaded by Noahcs at en.wikipedia.

File:Mount Juneau Alaska.jpg *Source*: http://en.wikipedia.org/w/index.php?title=File:Mount_Juneau_Alaska.jpg *License*: unknown *Contributors*: Aconcagua, Adam, Prankster

File:USACE Homer Spit Alaska.jpg *Source*: http://en.wikipedia.org/w/index.php?title=File:USACE_Homer_Spit_Alaska.jpg *License*: unknown *Contributors*: U.S. Army Corps of Engineers, photographer unknown

File:Kachcampus.jpg *Source*: http://en.wikipedia.org/w/index.php?title=File:Kachcampus.jpg *License*: unknown *Contributors*: Beeblebrox at en.wikipedia

File:Iditarod Ceremonial start in Anchorage, Alaska.jpg *Source*: http://en.wikipedia.org/w/index.php?title=File:Iditarod_Ceremonial_start_in_Anchorage,_Alaska.jpg *License*: unknown *Contributors*: Frank Kovalchek from Anchorage, Alaska, USA

File:Bow bow.jpg *Source*: http://en.wikipedia.org/w/index.php?title=File:Bow_bow.jpg *License*: unknown *Contributors*: User:Orinek7

File:2008-05-04 at 18-26-44-Forgetmenot-Flower.jpg *Source*: http://en.wikipedia.org/w/index.php?title=File:2008-05-04_at_18-26-44-Forgetmenot-Flower.jpg *License*: unknown *Contributors*: User:Wilder Kaiser

GNU Free Documentation License Version 1.2, November 2002 Copyright (C) 2000,2001,2002 Free Software Foundation, Inc. 59 Temple Place, Suite 330, Boston, MA 02111-1307 USA Everyone is permitted to copy and distribute verbatim copies of this license document, but changing it is not allowed.

0. PREAMBLE

The purpose of this License is to make a manual, textbook, or other functional and useful document "free" in the sense of freedom: to assure everyone the effective freedom to copy and redistribute it, with or without modifying it, either commercially or noncommercially. Secondarily, this License preserves for the author and publisher a way to get credit for their work, while not being considered responsible for modifications made by others. This License is a kind of "copyleft", which means that derivative works of the document must themselves be free in the same sense. It complements the GNU General Public License, which is a copyleft license designed for free software. We have designed this License in order to use it for manuals for free software, because free software needs free documentation: a free program should come with manuals providing the same freedoms that the software does. But this License is not limited to software manuals; it can be used for any textual work, regardless of subject matter or whether it is published as a printed book. We recommend this License principally for works whose purpose is instruction or reference.

1. APPLICABILITY AND DEFINITIONS

This License applies to any manual or other work, in any medium, that contains a notice placed by the copyright holder saying it can be distributed under the terms of this License. Such a notice grants a world-wide, royalty-free license, unlimited in duration, to use that work under the conditions stated herein. The "Document", below, refers to any such manual or work. Any member of the public is a licensee, and is addressed as "you". You accept the license if you copy, modify or distribute the work in a way requiring permission under copyright law. A "Modified Version" of the Document means any work containing the Document or a portion of it, either copied verbatim, or with modifications and/or translated into another language. A "Secondary Section" is a named appendix or a front-matter section of the Document that deals exclusively with the relationship of the publishers or authors of the Document to the Document's overall subject (or to related matters) and contains nothing that could fall directly within that overall subject. (Thus, if the Document is in part a textbook of mathematics, a Secondary Section may not explain any mathematics.) The relationship could be a matter of historical connection with the subject or with related matters, or of legal, commercial, philosophical, ethical or political position regarding them. The "Invariant Sections" are certain Secondary Sections whose titles are designated, as being those of Invariant Sections, in the notice that says that the Document is released under this License. If a section does not fit the above definition of Secondary then it is not allowed to be designated as Invariant. The Document may contain zero Invariant Sections. If the Document does not identify any Invariant Sections then there are none. The "Cover Texts" are certain short passages of text that are listed, as Front-Cover Texts or Back-Cover Texts, in the notice that says that the Document is released under this License. A Front-Cover Text may be at most 5 words, and a Back-Cover Text may be at most 25 words. A "Transparent" copy of the Document means a machine-readable copy, represented in a format whose specification is available to the general public, that is suitable for revising the document straightforwardly with generic text editors or (for images composed of pixels) generic paint programs or (for drawings) some widely available drawing editor, and that is suitable for input to text formatters or for automatic translation to a variety of formats suitable for input to text formatters. A copy made in an otherwise Transparent file format whose markup, or absence of markup, has been arranged to thwart or discourage subsequent modification by readers is not Transparent. An image format is not Transparent if used for any substantial amount of text. A copy that is not "Transparent" is called "Opaque". Examples of suitable formats for Transparent copies include plain ASCII without markup, Texinfo input format, LaTeX input format, SGML or XML using a publicly available DTD, and standard-conforming simple HTML, PostScript or PDF designed for human modification. Examples of transparent image formats include PNG, XCF and JPG. Opaque formats include proprietary formats that can be read and edited only by proprietary word processors, SGML or XML for which the DTD and/or processing tools are not generally available, and the machine-generated HTML, PostScript or PDF produced by some word processors for output purposes only. The "Title Page" means, for a printed book, the title page itself, plus such following pages as are needed to hold, legibly, the material this License requires to appear in the title page. For works in formats which do not have any title page as such, "Title Page" means the text near the most prominent appearance of the work's title, preceding the beginning of the body of the text. A section "Entitled XYZ" means a named subunit of the Document whose title either is precisely XYZ or contains XYZ in parentheses following text that translates XYZ in another language. (Here XYZ stands for a specific section name mentioned below, such as "Acknowledgements", "Dedications", "Endorsements", or "History".) To "Preserve the Title" of such a section when you modify the Document means that it remains a section "Entitled XYZ" according to this definition. The Document may include Warranty Disclaimers next to the notice which states that this License applies to the Document. These Warranty Disclaimers are considered to be included by reference in this License, but only as regards disclaiming warranties: any other implication that these Warranty Disclaimers may have is void and has no effect on the meaning of this License.

2. VERBATIM COPYING

You may copy and distribute the Document in any medium, either commercially or noncommercially, provided that this License, the copyright notices, and the license notice saying this License applies to the Document are reproduced in all copies, and that you add no other conditions whatsoever to those of this License. You may not use technical measures to obstruct or control the reading or further copying of the copies you make or distribute. However, you may accept compensation in exchange for copies. If you distribute a large enough number of copies you must also follow the conditions in section 3. You may also lend copies, under the same conditions stated above, and you may publicly display copies.

3. COPYING IN QUANTITY

If you publish printed copies (or copies in media that commonly have printed covers) of the Document, numbering more than 100, and the Document's license notice requires Cover Texts, you must enclose the copies in covers that carry, clearly and legibly, all these Cover Texts: Front-Cover Texts on the front cover, and Back-Cover Texts on the back cover. Both covers must also clearly and legibly identify you as the publisher of these copies. The front cover must present the full title with all words of the title equally prominent and visible. You may add other material on the covers in addition. Copying with changes limited to the covers, as long as they preserve the title of the Document and satisfy these conditions, can be treated as verbatim copying in other respects. If the required texts for either cover are too voluminous to fit legibly, you should put the first ones listed (as many as fit reasonably) on the actual cover, and continue the rest onto adjacent pages. If you publish or distribute Opaque copies of the Document numbering more than 100, you must either include a machine-readable Transparent copy along with each Opaque copy, or state in or with each Opaque copy a computer-network location from which the general network-using public has access to download using public-standard network protocols a complete Transparent copy of the Document, free of added material. If you use the latter option, you must take reasonably prudent steps, when you begin distribution of Opaque copies in quantity, to ensure that this Transparent copy will remain thus accessible at the stated location until at least one year after the last time you distribute an Opaque copy (directly or through your agents or retailers) of that edition to the public. It is requested, but not required, that you contact the authors of the Document well before redistributing any large number of copies, to give them a chance to provide you with an updated version of the Document.

4. MODIFICATIONS

You may copy and distribute a Modified Version of the Document under the conditions of sections 2 and 3 above, provided that you release the Modified Version under precisely this License, with the Modified Version filling the role of the Document, thus licensing distribution and modification of the Modified Version to whoever possesses a copy of it. In addition, you must do these things in the Modified Version: A. Use in the Title Page (and on the covers, if any) a title distinct from that of the Document, and from those of previous versions (which should, if there were any, be listed in the History section of the Document). You may use the same title as a previous version if the original publisher of that version gives permission. B. List on the Title Page, as authors, one or more persons or entities responsible for authorship of the modifications in the Modified Version, together with at least five of the principal authors of the Document (all of its principal authors, if it has fewer than five), unless they release you from this requirement. C. State on the Title page the name of the publisher of the Modified Version, as the publisher. D. Preserve all the copyright notices of the Document. E. Add an appropriate copyright notice for your modifications adjacent to the other copyright notices. F. Include, immediately after the copyright notices, a license notice giving the public permission to use the Modified Version under the terms of this License, in the form shown in the Addendum below. G. Preserve in that license notice the full lists of Invariant Sections and required Cover Texts given in the Document's license notice. H. Include an unaltered copy of this License. I. Preserve the section Entitled "History", Preserve its Title, and add to it an item stating at least the title, year, new authors, and publisher of the Modified Version as given on the Title Page. If there is no section Entitled "History" in the Document, create one stating the title, year, authors, and publisher of the Document as given on its Title Page, then add an item describing the Modified Version as stated in the previous sentence. J. Preserve the network location, if any, given in the Document for public access to a Transparent copy of the Document, and likewise the network locations given in the Document for previous versions it was based on. These may be placed in the "History" section. You may omit a network location for a work that was published at least four years before the Document itself, or if the original publisher of the version it refers to gives permission. K. For any section Entitled "Acknowledgements" or "Dedications", Preserve the Title of the section, and preserve in the section all the substance and tone of each of the contributor acknowledgements and/or dedications given therein. L. Preserve all the Invariant Sections of the Document, unaltered in their text and in their titles. Section numbers or the equivalent are not considered part of the section titles. M. Delete any section Entitled "Endorsements". Such a section may not be included in the Modified Version. N. Do not retitle any existing section to be Entitled "Endorsements" or to conflict in title with any Invariant Section. O. Preserve any Warranty Disclaimers. If the Modified Version includes new front-matter sections or appendices that qualify as Secondary Sections and contain no material copied from the Document, you may at your option designate some or all of these sections as invariant. To do this, add their titles to the list of Invariant Sections in the Modified Version's license notice. These titles must be distinct from any other section titles. You may add a section Entitled "Endorsements", provided it contains nothing but endorsements of your Modified Version by various parties--for example, statements of peer review or that the text has been approved by an organization as the authoritative definition of a standard. You may add a passage of up to five words as a Front-Cover Text, and a passage of up to 25 words as a Back-Cover Text, to the end of the list of Cover Texts in the Modified Version. Only one passage of Front-Cover Text and one of Back-Cover Text may be added by (or through arrangements made by) any one entity. If the Document already includes a cover text for the same cover, previously added by you or by arrangement made by the same entity you are acting on behalf of, you may not add another; but you may replace the old one, on explicit permission from the previous publisher that added the old one. The author(s) and publisher(s) of the Document do not by this License give permission to use their names for publicity for or to assert or imply endorsement of any Modified Version.

5. COMBINING DOCUMENTS

You may combine the Document with other documents released under this License, under the terms defined in section 4 above for modified versions, provided that you include in the combination all of the Invariant Sections of all of the original documents, unmodified, and list them all as Invariant Sections of your combined work in its license notice, and that you preserve all their Warranty Disclaimers. The combined work need only contain one copy of this License, and multiple identical Invariant Sections may be replaced with a single copy. If there are multiple Invariant Sections with the same name but different contents, make the title of each such section unique by adding at the end of it, in parentheses, the name of the original author or publisher of that section if known, or else a unique number. Make the same adjustment to the section titles in the list of Invariant Sections in the license notice of the combined work. In the combination, you must combine any sections Entitled "History" in the various original documents, forming one section Entitled "History"; likewise combine any sections Entitled "Acknowledgements", and any sections Entitled "Dedications". You must delete all sections Entitled "Endorsements".

6. COLLECTIONS OF DOCUMENTS

You may make a collection consisting of the Document and other documents released under this License, and replace the individual copies of this License in the various documents with a single copy that is included in the collection, provided that you follow the rules of this License for verbatim copying of each of the documents in all other respects. You may extract a single document from such a collection, and distribute it individually under this License, provided you insert a copy of this License into the extracted document, and follow this License in all other respects regarding verbatim copying of that document.

7. AGGREGATION WITH INDEPENDENT WORKS

A compilation of the Document or its derivatives with other separate and independent documents or works, in or on a volume of a storage or distribution medium, is called an "aggregate" if the copyright resulting from the compilation is not used to limit the legal rights of the compilation's users beyond what the individual works permit. When the Document is included in an aggregate, this License does not apply to the other works in the aggregate which are not themselves derivative works of the Document. If the Cover Text requirement of section 3 is applicable to these copies of the Document, then if the Document is less than one half of the entire aggregate, the Document's Cover Texts may be placed on covers that bracket the Document within the aggregate, or the electronic equivalent of covers if the Document is in electronic form. Otherwise they must appear on printed covers that bracket the whole aggregate.

8. TRANSLATION

Translation is considered a kind of modification, so you may distribute translations of the Document under the terms of section 4. Replacing Invariant Sections with translations requires special permission from their copyright holders, but you may include translations of some or all Invariant Sections in addition to the original versions of these Invariant Sections. You may include a translation of this License, and all the license notices in the Document, and any Warranty Disclaimers, provided that you also include the original English version of this License and the original versions of those notices and disclaimers. In case of a disagreement between the translation and the original version of this License or a notice or disclaimer, the original version will prevail. If a section in the Document is Entitled "Acknowledgements", "Dedications", or "History", the requirement (section 4) to Preserve its Title (section 1) will typically require changing the actual title.

9. TERMINATION

You may not copy, modify, sublicense, or distribute the Document except as expressly provided for under this License. Any other attempt to copy, modify, sublicense or distribute the Document is void, and will automatically terminate your rights under this License. However, parties who have received copies, or rights, from you under this License will not have their licenses terminated so long as such parties remain in full compliance.

10. FUTURE REVISIONS OF THIS LICENSE

The Free Software Foundation may publish new, revised versions of the GNU Free Documentation License from time to time. Such new versions will be similar in spirit to the present version, but may differ in detail to address new problems or concerns. See http://www.gnu.org/copyleft/. Each version of the License is given a distinguishing version number. If the Document specifies that a particular numbered version of this License "or any later version" applies to it, you have the option of following the terms and conditions either of that specified version or of any later version that has been published (not as a draft) by the Free Software Foundation. If the Document does not specify a version number of this License, you may choose any version ever published (not as a draft) by the Free Software Foundation. ADDENDUM: How to use this License for your documents To use this License in a document you have written, include a copy of the License in the document and put the following copyright and license notices just after the title page: Copyright (c) YEAR YOUR NAME. Permission is granted to copy, distribute and/or modify this document under the terms of the GNU Free Documentation License, Version 1.2 or any later version published by the Free Software Foundation; with no Invariant Sections, no Front-Cover Texts, and no Back-Cover Texts. A copy of the license is included in the section entitled "GNU Free Documentation License". If you have Invariant Sections, Front-Cover Texts and Back-Cover Texts, replace the "with...Texts." line with this: with the Invariant Sections being LIST THEIR TITLES, with the Front-Cover Texts being LIST, and with the Back-Cover Texts being LIST. If you have Invariant Sections without Cover Texts, or some other combination of the three, merge those two alternatives to suit the situation. If your document contains nontrivial examples of program code, we recommend releasing these examples in parallel under your choice of free software license, such as the GNU General Public License, to permit their use in free software.

MIX
Papier aus verantwortungsvollen Quellen
Paper from responsible sources
FSC® C105338

FSC
www.fsc.org

Printed by Books on Demand GmbH, Norderstedt / Germany